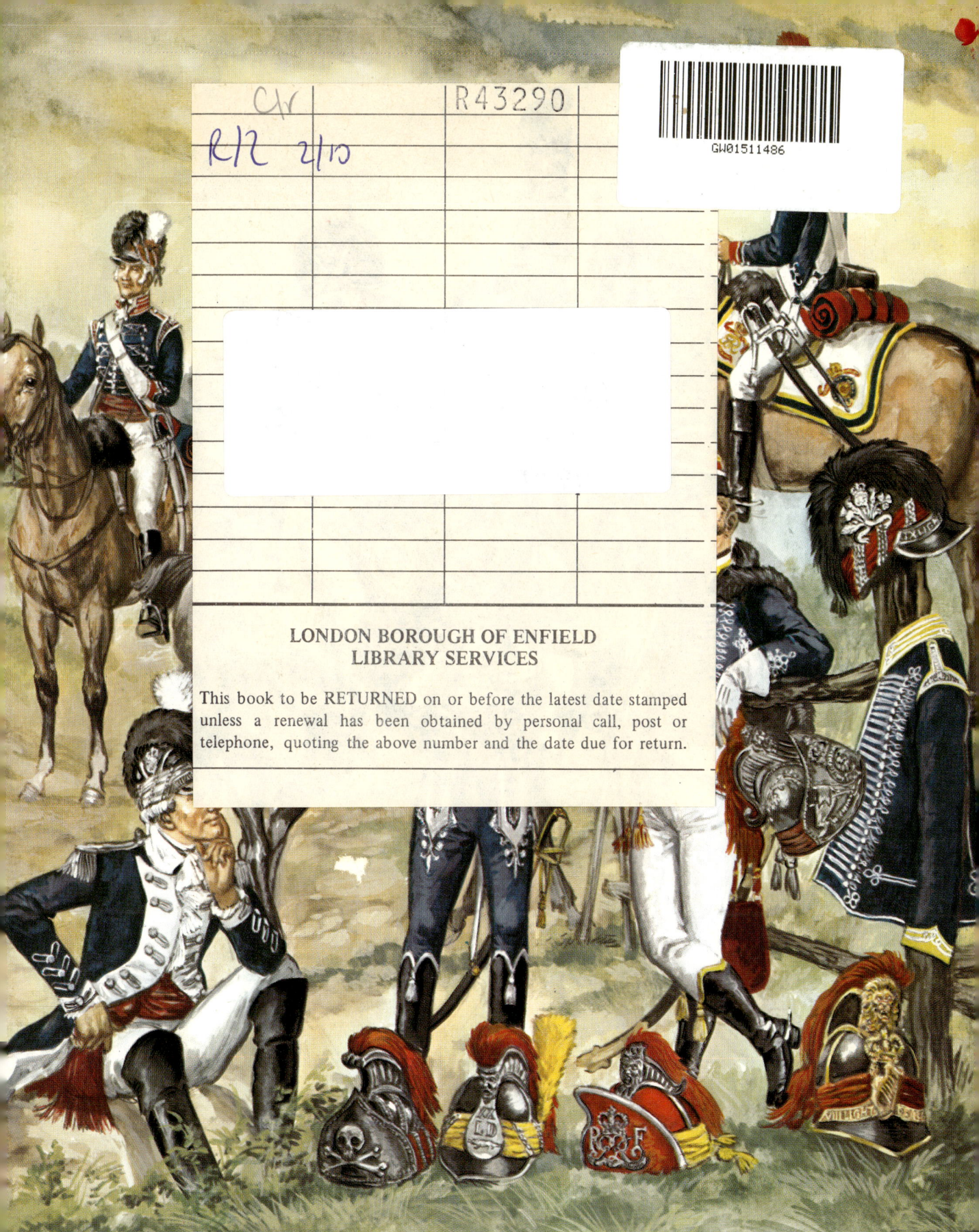

LONDON BOROUGH OF ENFIELD

Central Library
Cecil Road,
ENFIELD, Mddx

REFERENCE SERVICE

THE LACE WARS

ARMS and UNIFORMS

Other titles in this series:
Arms and Uniforms 1 – *Ancient Egypt in the 18th Century*
Arms and Uniforms 2 – *18th Century to the Present Day*
The Napoleonic Wars 1
The Napoleonic Wars 2
The First World War 1
The First World War 2
The Second World War 1
The Second World War 2
The Second World War 3
The Second World War 4
The Lace Wars I

THE LACE WARS

PART II
1700–1800: French, British and Prussian Cavalry and Artillery; Other countries: Infantry, Cavalry and Artillery

Liliane and Fred Funcken

WARD LOCK LIMITED . LONDON

Contents

Foreword	9
French Cavalry and Artillery	10
British Cavalry and Artillery	43
Cavalry and Artillery of Prussia, Saxony, Bavaria and other German States	64
Austria, Belgium, Italy and Spain	92
Russia and Sweden	128
Conclusion	154
Index	156

© Illustrations Casterman 1976
© Text Ward Lock 1977

First published in Great Britain in 1977 by Ward Lock Limited, 116 Baker Street, London W1M 2BB, a member of the Pentos Group.

All Rights Reserved. No part of this publication may be reproduced, stored in a retrieval system, or transmitted, in any form or by any means, electronic, mechanical, photocopying, recording, or otherwise, without the prior permission of the Copyright owners.

Text filmset in 'Monophoto' Baskerville by Servis Filmsetting Ltd, Manchester

Printed and bound in Belgium by Casterman S.A. Tournai

British Library Cataloguing in Publication Data

Funcken, Liliane
  The lace wars.
  Vol. 2: 1700–1800. – (Arms and uniforms).
  1. Uniforms, Military – History 2. Arms and armor – History
  I. Title II. Funcken, Fred III. Sharp, Richard IV. Series
  355.1'4'094   UC485.E/

ISBN 0-7063-5566-0

Foreword

In this second volume devoted to the eighteenth century we have brought together the uniforms of all the major European powers, as well as those of several minor states concerned in the various wars of the period.

To do justice to the exceptional variety and complexity of the subject, we have included nearly 1500 illustrations; even so we have had to restrict ourselves to a rather simplified study of some of the smaller armies, even to the point of omitting them altogether, as for instance in the case of the emergent American army, then involved in the struggle for independence. We hope to repair this particular omission in a forthcoming study of the uniforms of the United States. Uniforms of other countries are shown only at the most crucial stage of their military development – to cover the full course of the century in these countries would consume space that should be given to the four or five major powers.

Readers are likely to be particularly interested in the Russian army, which is here depicted in large-scale colour illustrations for the first time.

Anyone wanting to pursue this topic further will find it in all its intimidating majesty in the thirty volumes and 3,935 plates of Viskovatov's work, a few rare copies of which include some coloured illustrations. Equally imposing, if rather more practical, is Zvegintsov's abridged French version of Viskovatov's text and illustrations for the period under discussion, in three volumes: this also includes notes on the composition of the army, the history of its campaigns (with diagrams of the battles), and a chapter on manoeuvres and tactics. The same author also published *Flags and Standards of the Russian Army, from the Sixteenth Century to 1914*.

Our most sincere thanks are due to MM. Jacques Lesellier, Pierre Simon and Jacques Lekeu for all their kindness and their invaluable assistance; also to our good friend Eugène Leliepvre, official painter to the French Army, whose advice has been of the greatest value.

Liliane and Fred Funcken

FRENCH CAVALRY AND ARTILLERY

In the first volume of this study we concentrated largely on eighteenth-century France – the Royal Household and the infantry, with their impressive uniforms and their flags, weapons and drums. In this second volume we turn first to the equally distinguished French cavalry, then conclude our study of the French army with chapters on light troops and the artillery.

Cavalry

Under the Ancien Régime, the cavalry was always referred to as 'light cavalry', to distinguish it from the gendarmerie, considered then as the only 'heavy cavalry'. Not until 1791 did the 'light cavalry' regiments become recognized as cavalry proper.

The comprehensive organization of this branch of the service took place under Louis XIV's Colonel General of Cavalry, the famous Vicomte de Turenne.[1] The title of Colonel General survived only because of Turenne's powerful personality – in the infantry it had disappeared with the death of the Duc d'Épernon.[2] The rank of Colonel General of Cavalry, instituted by Louis XII, became a crown appointment in 1565, under Charles IX, and thenceforth involved exceptional privileges. The incumbent had absolute control of the discipline and administration of his troops; no operations could be undertaken, no commissions granted, without his authorization. Until 1789 the Colonel General was entitled to add to his coat of arms six standards bearing the fleur-de-lis, displayed in the form of a cross behind the escutcheon.

The title of Colonel was adopted by the cavalry in 1788 to replace the former designation of *mestre de camp*, the style previously given to the officer commanding a regiment. The title of Lieutenant Colonel had been used since the end of Louis XV's reign; in 1791 it referred to the officer commanding a cavalry squadron. In 1793 the rank was restyled *chef d'escadron* (squadron commander).

---

[1] Henri de la Tour d'Auvergne, Vicomte de Turenne, Marshal of France (1611–75).

[2] Bernard de Nogaret de la Valette, Duc d'Épernon (1592–1661) was appointed Colonel General of Infantry at the age of eighteen. When he died the colonels reappeared, abandoning their old title of *mestre de camp*, which could only be used when a Colonel General existed, and came back into use when first the son of the Regent (1721–30) and then the Prince de Condé (1780–8) were appointed to that rank. After Condé, the *mestres de camp* became Brigadiers until the decree of l Vendémiaire XII, which restored the title of Colonel.

FRANCE, CAVALRY (I)
1. Royal Carabineers, 1700. — 2. King's Cuirassiers, 1700. — 3. Villeroy Cavalry, 1724. — 4. Colonel General's Regiment, 1733. — 5. Rosen Cavalry, 1740.

The Regiments

The 'regular troops', i.e., those which did not form part of the Royal Household, were subdivided into three categories. The 'royal' regiments bore the names of the King, the Queen, the princes of the blood and the generals. At their head stood the three so-called headquarters regiments: those of the Colonel General,[1] the *Mestre de camp général* and the Commissary General. The 'noble' regiments were those commanded by officers other than the princes and the generals; their names changed frequently, as each regiment was always named after its current commander (or rather its proprietor). Finally there were the 'provincial' regiments raised by different foreign states.

The Sale of Commissions

Every commissioned rank was available for cash: a man would buy a regiment as he might buy an estate. In 1760 the King's minister Choiseul had attempted to put a stop to this custom: yet a young Colonel of noble birth might still say to his Lieutenant Colonel, 'You must recognize the difference between a man like you and a man like myself,' and receive the cutting answer: 'Yes, it takes forty thousand *écus* to make a man like you; men like me are made with forty years' service.'

In the reign of Louis XVI, Saint-Germain[2] resolutely attacked the problem of 'suppressing the buying and selling of all military commands', this practice being in his view 'the worst possible influence on the quality of the army. Money cannot buy either talent or merit, and the profession of arms requires a great deal of both'.

Saint-Germain's decree stipulated that each time a commission changed hands its price should be cut by a quarter of its original value, so that after four such changes each individual problem would have been solved. This progressive system was undoubtedly necessary; the State would have been quite incapable of buying up all the commissions itself at a single time. The 'owners' protested with the utmost vehemence against this assault on their 'divine rights'. But by the eve of the Revolution the project was well advanced.

To consider this decree democratic, would be a mistake. Its real effect was to purge the army of the very large number of sons of the wealthy middle class who, during the past two hundred years, had gradually supplanted the bankrupt aristocracy. The King was thus enabled to 'reward the impoverished nobility suitably for their distinguished services'. In 1781, moreover, the Comte d'Artois reinstituted the old qualification of nobility – four quarterings in the paternal line – for even the lowliest 2nd Lieutenant under his command. The

FRANCE, CAVALRY (II)
1. Royal German, 1754. – 2. King's Regiment, 1762. – 3. Royal German, training dress, 1750. – 4. Bourbon-Busset Regiment, in cloak, 1750. – 5. Colonel General's Trumpeter, 1758. – 6. Colonel General, 1758. In 1750 the uniform was the same but with epaulets and no lace on the buttonholes. – 7. *Mestre de camp général*. Until 1748 the colour of the horse's trappings was green. – 8. Commissary General, 1750. – 9. Dauphin's Regiment, travelling and manoeuvres dress, 1756. At the time of the Seven Years' War the tunic had fawn lace in imitation of the Prussians. The coat was in the saddle-bag with the cloak rolled up on top of it. – 10. Trumpeter in the Royal Regiments.
11. Breastplate (first model) with its shoulder-straps crossing at the back. – 12. Cartridge case with its twelve holes for cartridges, showing the double closing-system. – 13. Equipment belt. – 14. Regimental standards of the Colonel General's Regiment in 1753, 1773 and at the end of the monarchy.

---

[1] Founded in 1635, the Colonel General's Regiment enjoyed extensive privileges in many areas, from delivery of bread and forage to the choice of the best camp-sites or barrack quarters. It was always stationed on the right of the army. Its white standard, the *cornette blanche*, was dipped only to the King, the Princes of the Blood, the Colonel General of Cavalry and the Marshals of France; and it was itself saluted by all the other regimental standards. The rank of Cornet in the regiment, though only equivalent to that of junior Captain, was worth more than a regimental command. The Colonel's company was mounted on grey horses, the other eleven companies on black.

[2] See Vol. I, p. 40.

hopes aroused among the lower ranks by Saint-Germain's reforms were thus dashed, and the common people were left to proceed as best they could along the slow path of promotion from the ranks, a path that rarely led as high as the rank of major. It is easy to imagine how much fertile soil this must have provided for the sowing of revolutionary ideas.

Recruitment

In theory, only the Captain could legally enlist new recruits; in practice he delegated this duty to his subordinates, even to private soldiers. Officers going home on leave were also required to bring back a few recruits, and were ably assisted in this task by their relatives and friends who hunted out young men, 'sound in mind and body', to join the colours. Enticing posters were put up promising 'plenty of action, absolute freedom, forty sols a day travelling expenses, leave after eight years' service, and all sorts of other benefits.... Reward offered to anyone bringing suitable recruits to the colours.'

As one might expect from that last sentence, there were those for whom recruiting became a full-time profession. In time of peace it was not a particularly difficult one, but in wartime things were very different, and it was as much as the recruiting sergeants could do to make good the losses and keep their units up to strength. With their curled, powdered hair, acres of gold braid and often astounding line in headgear, the recruiting sergeants wooed their simple-minded audience with wine and promises. Any means of persuasion, even force, was good enough if it worked.

The recruits attracted by these fairground tactics were nowhere near the calibre of their peacetime equivalents, and their ranks were severely thinned by desertion before they ever reached their barracks. By Louis XVI's day, however, things had improved to the point where young recruits could be trusted to make their own way to barracks, with travelling expenses of three sous a league. It must be remembered, though, that this was an exceptionally peaceful period, apart from France's intervention in the American War of Independence, from 1776 to 1783.

Ransom

Extraordinary though it may seem to us nowadays, ransoming of prisoners was still very widespread throughout almost all the eighteenth century. It was a very long-established tradition, which allowed an army to recover the troops it had acquired with such difficulty. A treaty signed with England on 18 June 1743 fixed the price of an enlisted man at £4, a Sergeant fetched £10, a Captain £70, a Colonel £600, a Brigadier £900, a *mestre de camp* £1500, a Lieutenant General £15,000 and a Marshal £50,000.

It was a Captain's responsibility to ransom his own men; if he refused, another Captain had the right to buy up the abandoned troops for his own unit.

In 1780 another treaty with England modified the disparity between the different ranks: a soldier was now worth £25, and a Marshal, as the equivalent of sixty soldiers, fetched £1500. The Republic put a stop to this strange system of barter; under the new regime, prisoners were swapped on a strict rank-for-rank basis.

The ransom principle had at least one considerable advantage, from the humanitarian viewpoint. At the beginning of the century, by appealing to their natural greed, it helped to restrain the barbarity of the hussars and the irregular cavalry who had learnt during their fighting against the Turks to offer no quarter.

FRANCE, CAVALRY (III) 1740–86
A. 1740: 1. Colonel General.—2. *Mestre de camp général.*—3. Commissary General.—4. Royal.—5. King's.—6. Royal Foreign.—7. King's Cuirassiers.—8. Royal Cravate.—9. Royal Roussillon.—10. Royal Piedmontese.—11. Royal German.—12. Carabineers.—13. Royal Poland.—14. Queen's.—15. Dauphin's.—16. Dauphin's Foreign.—17. Brittany.—18. Anjou.—19. Berry.—20. Orléans.—21.

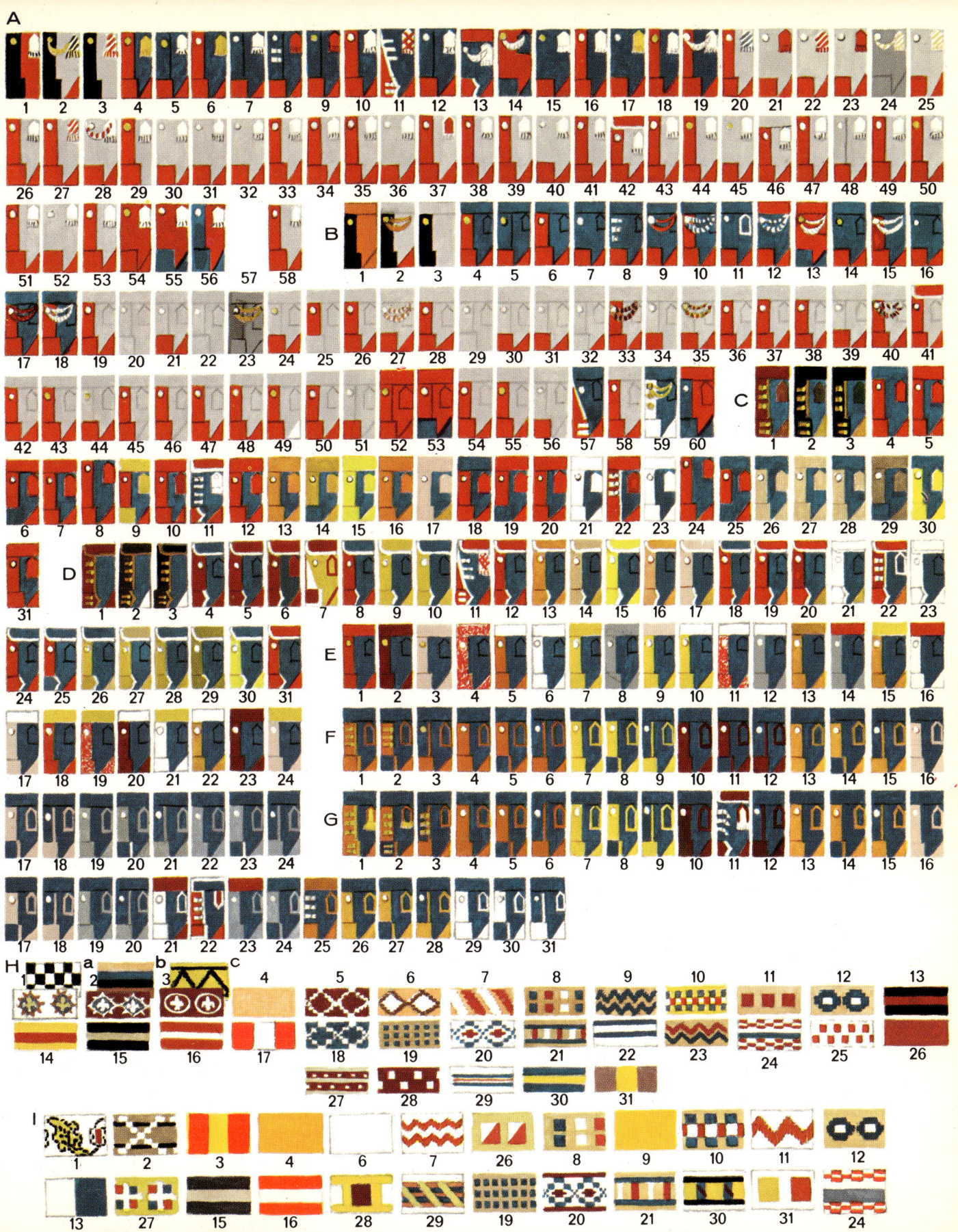

Condé.—22. Bourbon.—23. Clermont.—24. Conti.—25. Penthièvre.—26. Saint-Simon.—27. d'Ancezune.—28. Rohan.—29. Beaucaire.—30. Brancas.—31. Sabran.—32. Gesvres.—33. Chabrillant.—34. Chevalier de Rosen.—35. Saint-Aignan.—36. Grammont.—37. d'Andlau.—38. Fleury.—39. Sassenage.—40. Vogüé.—41. Vintimille.—42. Brissac.—43. Daumont.—44. Vasse.—45. La Ferronaye.—46. Randan.—47. d'Heudicourt.—48. Chépy.—49. Fiennes.—50. Lévy.—51. Barbanson.—52. Puysieulx.—53. Rosen.—54. Noailles.—55. Pons.—56. Fitz-James.—57. Rattky's Hussars.—58. d'Asfeld.—The hat was usually trimmed with silver, but gold occurs in a few regiments—the 1st, 2nd, 3rd, 4th, 5th, 9th, 14th, 15th, 18th, 29th, 49th, 52nd and 54th. The Royal German Regiment wore a fur cap, in black, with a red *flamme*.

B. 1757: 1. Colonel General.—2. *Mestre de camp*.—3. Commissary General.—4. Royal.—5. King's.—6. Foreign.—7. Cuirassiers.—8. Royal Cravate.—9. Royal Roussillon.—10. Royal Piedmontese.—11. Carabineers.—12. Royal Poland.—13. Queen's.—14. Dauphin's.—15. Dauphin's Foreign.—16. Burgundy.—17. Aquitaine.—18. Berry.—19. Orléans.—20. Condé.—21. Bourbon.—22. Clermont.—23. Conti.—24. Penthièvre.—25. Archiac.—26. Poly Saint-Thiébault.—27. Lusignan.—28. Marcieux.—29. Des Salles.—30. Talleyrand.—31. Clermont-Tonnerre.—32. Chabrillant.—33. d'Egmont.—34. Beauvilliers.—35. Grammont.—36. Bourbon-Busset.—37. Viefville.—38. Maugiron.—39. Saint-Jal.—40. Fumel.—41. Rochefoucauld-Langeac.—42. de Vienne.—43. Bussy-Lameth.—44. Crussol.—45. Fleury.—46. Léhoncourt.—47. Bellefonds.—48. Dampierre.—49. Henrichemont.—50. Moustiers.—51. Saluces.—52. Ayen.—53. Harcourt.—54. Descars.—55. Moncalm.—56. Bezons.—57. Royal German.—58. Württemberg.—59. Nassau-Sarrebruck.—60. Fitz-James.—In general the braid on the hat matched the metal of the buttons, though it should be noted that the 17th Regiment had gold braid, as did the 23rd, 29th and 48th. The 43rd, 44th and 49th, on the other hand, had silver braid despite their gold buttons. The Royal German (fig. 57) wore black fur caps with a red *flamme*. The Württemberg, Nassau-Sarrebruck and Royal Poland Regiments, as well as the King's Cuirassiers and, later, the Penthièvre and Orléans Regiments, also adopted this headgear.

C. 1762: 1. Colonel General.—2. *Mestre de camp*.—3. Commissary General.—4. Royal.—5. King's.—6. Foreign.—7. Cuirassiers.—8. Royal Cravate.—9. Royal Roussillon.—10. Royal Piedmontese.—11. Royal German.—12. Royal Poland.—13. Lorraine.—14. Picardy.—15. Champagne.—16. Navarre.—17. Normandy.—18. Queen's.—19. Dauphin's.—20. Burgundy.—21. Berry.—22. Carabineers.—23. Artois.—24. Orléans.—25. Chartres.—26. Condé.—27. Bourbon.—28. Clermont.—29. Conti.—30. Penthièvre.—31. Noailles.—The difference between the *mestre de camp general's* and the Commissary General's Regiments was in the orange-yellow braid on the buttonholes of the lower lapel and pockets. The first three regiments also had this orange-yellow braid on their hats; all the others had silver.

D. 1767: 1. Colonel General.—2. *Mestre de camp*.—3. Commissary General.—4. Royal.—5. King's.—6. Foreign.—7. Cuirassiers.—8. Royal Cravate.—9. Royal Roussillon.—10. Royal Piedmontese.—11. Royal German.—12. Royal Poland.—13. Lorraine.—14. Picardy.—15. Champagne.—16. Navarre.—17. Normandy.—18. Queen's.—19. Dauphin's.—20. Burgundy.—21. Berry.—22. Carabineers.—23. d'Artois.—24. Orléans.—25. Chartres.—26. Condé. 27. Bourbon.—28. Clermont.—29. Conti.—30. Penthièvre.—31. Noailles.—Only the first three regiments had gold braid on their tricorn hats. According to regulations the 11th (Royal German) Regiment wore black fur caps with white cord, *raquettes* and plume.

E. 1776: 1. Colonel General.—2. *Mestre de camp*.—3. Commissary General.—4. Royal.—5. King's.—6. Foreign.—7. King's Cuirassiers.—8. Royal Cravate.—9. Royal Roussillon.—10. Royal Piedmontese.—11. Royal German.—12. Royal Poland.—13. Lorraine.—14. Picardy.—15. Champagne.—16. Navarre.—17. Normandy.—18. Queen's.—19. Dauphin's.—20. Burgundy.—21. Berry.—22. Carabineers.—23. d'Artois.—24. Orléans.

F. 1779: 1. Colonel General.—2. *Mestre de camp général*.—3. Commissary General.—4. Royal.—5. King's.—6. Royal Foreign.—7. Cuirassiers.—8. Royal Cravate.—9. Royal Roussillon.—10. Royal Piedmontese.—11. Royal German.—12. Royal Poland.—13. Royal Lorraine.—14. Royal Picardy.—15. Royal Champagne.—16. Royal Navarre.—17. Royal Normandy.—18. Queen's.—19. Dauphin's.—20. Burgundy.—21. Berry.—22. Carabineers.—23. Artois.—24. Orléans.—Hats have now lost their gold or silver braid.

G. 1786: 1. Colonel General.—2. *Mestre de camp général*.—3. Commissary General.—4. Royal.—5. King's.—6. Royal Foreign.—7. Cuirassiers.—8. Royal Cravate.—9. Royal Roussillon.—10. Royal Piedmontese.—11. Royal German.—12. Royal Poland.—13. Royal Lorraine.—14. Royal Picardy.—15. Royal Champagne.—16. Royal Navarre.—17. Royal Normandy.—18. Queen's.—19. Dauphin's.—20. Burgundy.—21. Berry.—22. Carabineers.—23. Artois.—24. Orléans.—25. Nassau-Sarrebruck.—26. Orléanais.—27. Évêchés.—28. Franche-Comté.—29. Septimanie.—30. Quercy.—31. La Marche.—Facings were decorated with a blue fleur-de-lis. Only the 11th (Royal German) Regiment was permitted to wear the fur cap. The regiments numbered 27 to 31 had only a brief existence from 1784 until 1788, and their braids as shown here is only an attempt at reconstruction (see series I). The trimmings worn by the 25th Regiment is unknown—probably there never was one, the regiment having existed on paper only.

H. Saddlecloth trimming: a, b, c: 1st, 2nd and 3rd Regiments in 1750. At this time the saddlecloth and holsters had to be in blue cloth, but the princes' regiments (Bourbon, Orléans and Penthièvre) had red, while the Clermont, Condé and Conti regiments had buff. The same remained true in 1762 and 1767.—The braid numbered 1 to 31 is that of the 1762 regiments (see series C above); those numbered 4 to 12, 18, 19, 21, 24, 26, 27, 30 and 31 existed in the same form in 1750.

I. Braid in 1786: Their numbering sequence corresponds to that given in series G above. Those of regiments not shown here were identical to the 1762 series (H). The background colour of saddlecloth and holsters was officially royal blue, but again red was preferred in the first three, so-called headquarters regiments.

Organization of the Cavalry

In Turenne's day, the cavalry lagged behind the infantry as far as organization was concerned. In 1654 the idea of the squadron was introduced, each squadron consisting of two companies of forty-six men (usually of noble birth). Not until 1668 did one of the first cavalry Brigadiers, Fourilles, introduce any real organization: he formalized cavalry tactics, and divided the sixty-six regiments into squadrons of four companies each.

By the time Louis XIV died in 1715 he had reduced the number of regiments to twenty-four; but by 1724 there were fifty-nine of them, each comprising two squadrons of four companies. Each company counted thirty-two men, including officers.

The army reform of 1749 considerably reduced the numbers, but a statute passed in 1755 increased the strength of each company to forty men. And at the end of Louis XV's reign (1715–74) a further reorganization in 1772 changed the make-up of each regiment from four squadrons (each of two companies of fifty-four men) to three (each of four companies of thirty-six men).

In 1776 the Comte de Saint-Germain reduced the cavalry to twenty-four regiments, only to increase it again to thirty-one, ten years later; each regiment now counted five squadrons, including a reserve, of ninety-two men each, with the subdivision into companies being abolished. One exception was the Regiment of Carabineers, which comprised five brigades of two squadrons each, a total strength of 1160 men.

[1] Étienne-François, Duc de Choiseul (1719–85). After a distinguished military career, he turned to diplomacy, and became Secretary of State for Foreign Affairs, which position he abandoned for the Ministry of War (1761–70) and then for the Ministry of the Marine (1761–6). He introduced wide-ranging reforms in both these areas. A tireless worker, he was at the same time, the guiding force of French politics.
[2] 'Airs' in riding-school language refers to artificial manoeuvres and more or less rhythmic movements.

Horsemanship

With the reorganization of the cavalry and the introduction of formalized tactics under Choiseul,[1] in the reign of Louis XV, came the first riding schools, set up to ensure uniform standards of instruction. Schools opened in 1763 at Saumur (this one was reserved for the carabineers), at Douai, Metz, Besançon and Cambrai (this one reserved for the dragoons). A sixth was started at la Flèche in 1764. In 1766 Choiseul decided to retain only Saumur, and each regiment was ordered to send a detachment there. This famous school, together with the riding-schools at Saint-Germain and Versailles, provided the army with its riding instructors. Among them special mention should be made of the father of the old French *haute école*, François Robichon de la Guérinière, equerry to Louis XV.

Parrocel's exquisite engravings, showing the 'airs on the ground' (*passage, galopade, piaffer*) and the 'airs above the ground'[2] (*pesade, mézoir, courbette, croupade* and *capriole*) have an elegance which ensures that they are still in great demand even today. La Guérinière's *L'École de cavalerie* (1712) and *Éléments de cavalerie* were followed by Comte Drummont de Belfort's *Essai sur la cavalerie légère*, the first drill manual for cavalry, published in 1748, which first publicized the ideas of military horsemanship, later perfected by Seydlitz and Zieten in Prussia.

The ferment of the Revolution closed most of the riding schools and academies. Saumur itself closed its doors in 1790, the result of the departure of the carabineers, the statute of 1788, and the drying up of the funds allocated to it by the Treasury.

Tactics

At the end of the seventeenth century, tactics were still ill defined, usually consisting of a prolonged exchange of shots followed by a charge at the trot. Turenne favoured an informal, free-for-all style of charge; Condé the mass charge, at the gallop and with drawn swords.

The disappearance from the battlefield of the formidable pikemen, who were increasingly replaced by musketeers, encouraged the cavalry to risk the violent frontal attack; the slow loading and erratic functioning of the musket, which was uncertain in wet weather, were an added incentive to more frequent cavalry charges. So great did the horsemen's confidence become that neither the introduction of the percussion cap nor the invention of the *modern bullet* could deter them from their favourite tactic in the course of the following century. It was not until the appalling slaughter of 1914 was it admitted that the cavalry no longer had a part to play on the field of battle.

Standards

Under Louis XV cavalry standards bore the royal motto *Nec pluribus impar*[1] below a shining sun, while the reverse side bore the personal arms of the *mestre de camp*.[2]

A statute of 1689 allocated two standards to each squadron, each entrusted to an officer known as a 'cornet'. This rank was abolished in 1762, and replaced by that of ensign, except in the Colonel General's Regiment, which kept its cornets and its distinctive *cornette blanche*.[3] The title of ensign itself became obsolete in 1772.

In 1760 the number of standards per squadron was reduced to one, but a statute of 1784 restored the status quo. As before, the standards were usually in the personal colours of the *mestre de camp*, with his coat of arms on the reverse and the royal sun on the front. In the regiments officially designated 'royal', however, the background colour was always blue, and the royal sun was augmented with a fleur-de-lis in each corner. The reverse sometimes showed a pattern of fleurs-de-lis (King's Regiment, Royal Cuirassiers, Royal Cravate, Royal Piedmontese and Royal Poland), sometimes simply repeated the same pattern as the front (Royal Regiment, Royal Rousillon, Royal German, Royal Carabineers).

FRANCE, CAVALRY (IV)
Ranks in 1786 (gold replaced silver when gold buttons were worn): 1. *Mestre de camp* commanding.—2. Assistant *mestre de camp*.—3. Subsidiary *mestre de camp*.—4. Officer of the rank of Brigadier. If the regiment to which he belonged had gilded buttons the star would be silver and the epaulet itself gold.—5. Major. The Captain commanding wore an epaulet similar to the Major's, but on the left shoulder only, whereas his superiors had one on each shoulder.—On the left shoulder only: 6. Captain (second-in-command).—7. Reserve Captain.—8. First Lieutenant.—9. Second-Lieutenant.—10. Sub-Lieutenant.—11. Reserve Sub-Lieutenant.—12. Ensign.—13. Company Sergeant Major.—14. Gentleman trooper.—15. Sergeant.—16. Corporal clerk. In 1791 the same stripes were given to quartermaster-corporals, with an additional gold or silver stripe sewn diagonally above the angle of the arm.—17. Sergeant.—18. Corporal.—19. Lance-Corporal.—20. Barber-Surgeon.—The Blacksmith was distinguishable by a 23-mm-wide horseshoe in white thread above the elbow.
1. Trumpeter of the Bourbon Cavalry in 1760.—2. Sergeant of the Normandy Regiment in 1767.—3. N.C.O. of the Royal Cravate Regiment in 1776. His lapels are fastened all the way down.—4. Royal Normandy in 1790–5. Colonel General's Regiment in 1786.—6. *Mestre de camp général's* Regiment in 1786.—7. Commissary General's Regiment in 1786. The leather gloves, also visible in fig. 6, were worn only on parade. There are a few minor differences of colour noticeable in comparison to the previous period.—8. King's Cuirassiers (7th Regiment) in 1785.—9. The same, seen from the rear.—10. Trumpeter of the Commissary General's Regiment, 1786.—11. Trumpeter of the Berry Cavalry, 1786. His dress is as prescribed for all the cavalry regiments except the headquarters, Queen's and princes' regiments.

---

[1] 'Not unequal to several [suns]', i.e. 'superior to all', a motto inherited from Louis XIV, who had taken the sun as his personal emblem.
[2] And not, as often supposed, those of the Colonel in Chief, since there were no colonels in the cavalry until 1788.
[3] See above, in the chapter on 'The Regiments'.

Around 1730 the standards were square in shape, measuring about 60 or 65 centimetres each side. By 1775 the shape had been changed slightly, giving a height of 55 centimetres and a breadth of 65. Each was carried on a staff in the form of a tilting lance, painted in the same colour as the background of the standard.

The Cuirassiers

This regiment – the 7th – had worn the *cuirass* (breastplate and backplate) since the days of Louis XIV. Known under Louis XV as the Royal Cuirassiers, it was reduced to two squadrons in 1745 and renamed the King's Cuirassiers. Saint-Germain's reforms of 1774 left it with six squad-

FRANCE, CAVALRY (V)
1. Trooper of the Comte de Provence's Carabineers, 1758. This regiment, said to be 'worth five ordinary regiments', was known as the 'corps of carabineers'.—2. Carabineer trumpeter in 1786. The same red coat was worn by the trumpeters in the 1st, 3rd, 18th, 22nd and 24th Regiments, while the 2nd and 23rd wore green. For all other regiments, see fig. 7.—3. 1st Regiment of Monsieur's Carabineers in 1788. The lower part of the holster-covers, not shown here, was rectangular in shape and had the same braid as the upper part.—3a. Cuff design for the 2nd Regiment, 1788.—4. Carabineer of the 1st Regiment, 1791. It was in this year that the carabineers took to wearing the fur cap of the elite regiments, and that the two carabineer regiments took precedence over the rest of the cavalry.—4a. Cuff, 1st Regiment.—4b. Cuff, 2nd Regiment.—5. 8th Regiment of cavalry, dressed for duty on foot (without cuirass) in 1792.—6. 4th Regiment in 1792. By this time only the 8th Regiment (former cuirassiers) wore the whole cuirass, with breastplate and backplate the first four regiments, equipped with cuirasses in September 1792, in fact had only the breastplate, as did the 5th, 6th and 7th Regiments who later followed their example.—7. Trumpeter of the 19th Regiment in 1789.—8. Trumpeter of the 14th Regiment (third series) in 1791. His sleeves have chevrons trimmed in the royal livery. —9. Horse's trappings in 1791. The embroidery was in the appropriate colour for each series of regiments (see below). It seems certain, however, that the model in use at this time was the 'modified' 1786 version (fig. 9a), followed by the 1791 model, which had a square-cornered portmanteau, (fig. 9b), between about 1800 and 1802.

Provisional Statute of 1 April 1791
From now on, regiments had no titles, simply numbers. The 24 regiments were divided into 4 series, each with its own distinguishing colour. Each series was in turn divided into two sub-series, one with horizontal pockets and one with vertical. As it was necessary to be able to distinguish between the three regiments of each sub-series, the first was given collar, cuffs and cuff-straps in the distinctive colour, the second cuffs only, and the third collar and cuff-straps only.

Thus we have illustrated the first series only (distinguishing colour scarlet): 1. 1st Regiment (formerly Colonel General's) —2. 2nd Regiment (formerly Royal); the *Mestre de camp général's* Regiment had vanished after its disgrace and loss of privileges for its part in the Nancy uprisings of 1790.—3. 3rd Regiment (formerly Commissary General's).—4. 4th Regiment (formerly Queen's).—5. 5th Regiment (formerly Royal Poland).—6. 6th Regiment (formerly King's).

Details of the other eighteen regiments can easily be worked out by replacing the scarlet with pale yellow for the second series, crimson for the third and pink for the fourth. Using the same system of numbering as in the illustration, the regiments are:

Second series (pale yellow)
1. 7th Regiment (formerly Royal Foreign)
2. 8th Regiment (formerly Cuirassiers)
3. 9th Regiment (formerly Artois)
4. 10th Regiment (formerly Royal Cravate)
5. 11th Regiment (formerly Royal Roussillon)
6. 12th Regiment (formerly Dauphin's)

Third series (crimson)
1. 13th Regiment (formerly Orléans)
2. 14th Regiment (formerly Royal Piedmontese)
3. 15th Regiment (formerly Royal German)
4. 16th Regiment (formerly Royal Lorraine)
5. 17th Regiment (formerly Royal Burgundy)
6. 18th Regiment (formerly Berry)

Fourth series (pink)
1. 19th Regiment (formerly Royal Normandy)
2. 20th Regiment (formerly Royal Champagne)
3. 21st Regiment (formerly Royal Picardy)
4. 22nd Regiment (formerly Royal Navarre)
5. 23rd Regiment (formerly Royal Guyenne)
6. 24th Regiment.

When the 15th (Royal German) Regiment emigrated at the beginning of the year 1792, the nine higher-numbered regiments moved up a place to become the 15th to 23rd.

rons – four of cuirassiers, one of light horse (the 5th Squadron) and one reserve (the 6th). After the old regimental names were suppressed by the Law of 1 January 1791 the regiment became known simply as the 8th, and was the only one to retain the full cuirass until 1802. The breastplate alone was in theory worn by all cavalry regiments, but there was little enthusiasm for it, and it was often left off, contrary to regulations. This item of armour, made of browned iron and fastened over the leather jerkin by two straps crossing at the back, was officially discarded in 1767. Louis XVI tried to reintroduce it, but without success. Officers were still supposed to wear the full cuirass (made in their case of polished metal), but rarely did so, despite the example set by their sovereign, Louis XV, virtually all of whose surviving portraits show him wearing the cuirass.

The Carabineers

In 1693 the various carabineer companies, elite units of which each regiment had one, were reconstituted as a single regiment, the Royal Carabineers. Between 1715 and 1734, however, these specialists were again to be found in the cavalry regiments, in the ratio of four per company.

In 1758 the Comte de Provence – the King's younger brother, the future Louis XVIII, traditionally known as 'Monsieur' – assumed command of the Royal Carabineers, who now became 'Monsieur's Carabineer Corps'. In 1775 the corps was reduced to 1,200 horse, and in 1788 it was split into two regiments, who led the cavalry under the Revolution in 1791.

The characteristic weapon used by these elite troops was the rifled carbine, firing buckshot or balls which were tapped into the barrel of the gun with a mallet, allowing the rifling to grip them. These bullets acquired far greater velocity and accuracy than those fired from ordinary cavalry carbines.

The Dragoons

The name 'dragoon' is of controversial origin. Some trace it back to the dragon standards of Rome, and the *draconarii* who carried them; others to some German origin, or to the officers commanding the first mounted infantry raised by the Duc de Brissac – a force modelled on the Italian mounted arquebusiers, to please Marie de Medici. The dragons embroidered on their banners gave rise to the phrase 'raising dragon'. The pro-Italian sentiment of the Renaissance makes this version entirely plausible; but with Latin coming back into favour as it did at that period one cannot altogether discount the '*draconarii*' version. British

FRANCE, DRAGOONS (I)
A: 1. Dragoon, early years of the eighteenth century. — 2. Dragoon of the Colonel General's Regiment, 1720. A black leather tool-case was later introduced, its shape varying according to whether the dragoon was equipped with a shovel, a pick, an axe or a bill-hook. When natural leather harness was introduced at the end of the century, these tool cases were changed to match the new type of leather in use. — 3. Dragoon of Harcourt's Regiment in 1750. Gaiters are no longer buckled but laced instead. — 4. Foot dragoon of the Royal Regiments in 1750. Note the special bayonet (and see also the fourth page of dragoon illustrations).
B. 1733: 1. 1st, Colonel General's. — 2. 2nd, *Mestre de camp général's*. — 3. 3rd, Royal. — 4. 4th, Queen's. — 5. 5th, Dauphin's. — 6. 6th Orléans. — 7. 7th, Condé. — 8. 8th, Beauffremont. — 9. 9th, d'Armenonville. — 10. 10th, Vibraye. — 11. 11th, Saint-Mesme. — 12. 12th, Harcourt's. — 13. 13th, Nicolai. — 14. 14th, La Suze. — 15. 15th, Languedoc.
C. 1750: 1. Colonel General's. — 2. *Mestre de camp général's*. — 3. Royal. — 4. King's. — 5. Queen's. — 6. Dauphin's. — 7. Orléans. — 8. Beauffremont. — 9. Aubigné. — 10. Caraman. — 11. La Feronnaye. — 12. Harcourt. — 13. d'Apchon. — 14. Thyanges. — 15. Marbeuf. — 16. Languedoc.

Red coat lapels are also found in the 12th Regiment; and all-red coat and tunic in the 13th. Saddlecloth and holster-covers of the 1st Regiment are also shown with a blue background. Hats were always trimmed with silver, except in the 15th (Marbeuf) Regiment where gold braid was used. Coloured breeches gave place to leather ones during the same year, as the regulation of 1 May 1750 was gradually put into force.

authors connect the name with a short musket originally issued to the first British dragoons.

At the end of the nineteenth century Henri Choppin[1] suggested yet another very tempting solution, which referred back to Guillaume de Gomiecourt, called 'The Dragon', who was famous for his campaigns against the English in the twelfth century, and to his successor Raoul ('Dragon') de Gomiecourt who later raised a light troop of dragoons to fight on horseback or on foot.

Whatever the origin of their own name, the dragoons undoubtedly left a sinister verbal legacy of their own – the word *dragonnade*, coined to describe their infamous persecution of the French protestants in 1685,[2] and 'dragoon' itself came to be a verb meaning 'to bully, humiliate, insult and pillage'. But this reflects the rough behaviour of seventeenth-century soldiery in general. By the time of the Regency and the early years of the eighteenth century things had changed considerably for the better.

The development of the dragoon regiments and their uniforms is clearly shown in the accompanying illustrations. It is however as well to stress the abrupt change of 21 December 1762, when the green coat and the famous 'Schomberg' helmet made their appearance. This headgear, with its variants, was a tremendous success, the dragoons being extremely reluctant to remove it, even in church. Choiseul had to intervene personally in 1765: 'His Majesty, considering that dragoons when not on duty at church should be regarded as private persons and should therefore conform to the normal requirements of decency and courtesy, therefore requires that the dragoons should be obliged to remove their caps and go bare-headed in church as do the rest of the faithful.'

Dragoon officers, more determinedly eccentric than those of other branches of the service, wore their helmets tilted forwards, almost over their eyes; in 1782 they flatly refused to wear peaked helmets. One example of their originality, from among many: the Dauphin, son of Louis XV and Colonel General of Dragoons, is said to have had a crest made for his helmet out of women's hair.

Naturally, these regiments, in which young men of wealthy and titled families could serve without loss of status, commanded record prices – on occasion as much as £100,000.[3]

FRANCE, DRAGOONS (II)

A: 1. Dragoon of Apchon's Regiment in 1750. Despite the impression given by countless pictures of dragoons, the cap or *chaperon* coexisted with the hat, which appeared as early as 1696. In principle, the cap was to be worn only for royal inspections, or when the commander ordered it, or to go foraging. Even so, dragoons seem often to have worn this headgear in the field. The Colonel General's Regiment, on the other hand, is reputed to have worn its caps only for royal inspections. For normal inspections it was customary to place the cap on the *horse's* head! – 2. Dragoon of the commanding officer's company, 1st (Colonel General's) Regiment, in 1763. The company was identifiable by its white helmet-crests and grey horses. Note the rolled-up *pokalem* (see fig. 3) above the ammunition-pouch; it was normally carried in a white cylindrical case. The saddle shown here is the old-style one, the new version being in natural leather. – 3. The *pokalem*. – 4. Distinctive saddle-cloth decoration of the *Mestre de camp général's* Regiment. – 5. Schomberg's Regiment (17th and last in the series shown at C on this page). The exaggeratedly short boots are very typical of the period and simply reflect one of the numerous eccentricities of contemporary fashion. For service on foot, black cloth gaiters were worn. It should be added that Schomberg's Regiment had horizontal pockets, and that their saddlecloth was white with a double black edging.
B. 1757: 1. Colonel General's. – 2. *Mestre de camp général's*. – 3. Royal. – 4. King's. – 5. Queen's. – 6. Dauphin's. – 7. Orléans. – 8. Beauffremont. – 9. Aubigné. – 10. Caraman. – 11. La Ferronnaye. – 12. Harcourt-Beuvron. – 13. Apchon. – 14. Thyanges. – 15. Marbeuf. – 16. Languedoc. – The saddlecloth was identical to that in use in 1750, shown on the preceding page. The fringed epaulet whose colours matched the lace is sometimes shown on the left shoulder, but was in fact worn on the right from 1757, while the left shoulder displayed a shoulder strap in the same colour as the coat, bordered with white. A slightly later manuscript (1761) shows the fringed epaulets on the left shoulder, their colours no longer matching the saddlecloth. The hearts shown on the coat facings in our drawings are also taken from this source.
C. 1762: 1. Colonel General's. – 2. *Mestre de camp général's*. – 3. Royal. – 4. King's. – 5. Queen's. – 6. Dauphin's. – 7. Orléans. – 8. Beauffremont. – 9. Choiseul. – 10. d'Autichamps. – 11. Chabot. – 12. Coigny. – 13. Nicolai. – 14. Chapt. – 15. Chabrillant. – 16. Languedoc. – The 17th Regiment from this period is shown above at A (fig. 5).

---

[1] *La Cavalerie française* (1893).
[2] After the repeal of the Edict of Nantes. The protestants were made solely responsible for providing billets for the dragoons, and the military were encouraged to have a good time. These 'missionaries with boots on' obtained tens of thousands of conversions in a few months, especially in Poitou, Languedoc and Béarn.
[3] By way of comparison, an ordinary infantry regiment cost £20,000.

The Light Horse

The six regiments of light horse had a brief life. They originated in 1779, when each cavalry regiment had a fifth squadron composed of light horse; they disappeared in 1784. The 1st Regiment became first the Orléans Cavalry, then in 1788 the Royal Guyenne, then the 24th and finally the 22nd Regiment of Cavalry. The other five became the Évêchés, Franche-Comté, Septimanie, Quercy and La Marche Regiments of Cavalry. In 1788 all were converted to mounted chausseurs.

FRANCE, DRAGOONS (III)

A: 1. Drummer in a royal regiment, 1750.—2. Drummer, Orléans Regiment, 1770.—3. Guidon of the Languedoc Regiment, 1786. The silver scroll above the sun carries the famous inscription *Nec pluribus impar*.—4. Dragoon of Belsunce's Regiment in 1779. The veteran's medallion, in the regimental colour and embellished with copper, had been introduced in 1771, and was worn after a soldier's third engagement. The first two actions were denoted by one and two reversed chevrons respectively (see Hussars). These insignia were worn throughout the army.—5. Trumpeter of the 3rd Regiment of mounted chasseurs, 1779 (see series C, below).

Guidons: 6. Royal Dragoon Regiment.—7. Dauphin's Regiment—the only difference between this and the previous emblem is that here fleurs-de-lis alternate with dolphins. Since the cession of the Dauphiné to France in 1349, the eldest sons of the monarch had traditionally borne the title of Dauphin ('dolphin').—8. Orléans Dragoons.—9. 1st Company, Colonel General's Regiment. The other companies had the same guidon, but with a crimson background. These guidons existed as such during most of the eighteenth century. The reader who finds special interest in the few flags and standards we have shown is referred to P. Charrié's illustrations in *Le Plumet*, Rigo's very fine and scholarly sequence of plates (see Vol. I, p. 48), and the series done by M. Fouré of St-Cloud.

B. 1776: 1. Colonel General.—2. *Mestre de camp général*.—3. Royal.—4. King's.—5. Queen's.—6. Dauphin's.—7. Monsieur's.—8. Comte d'Artois's.—9. Orléans.—10. Chartres.—11. Condé.—12. Bourbon.—13. Conti.—14. Penthièvre.—15. Boufflers.—16. Lorraine.—17. Custine.—18. La Rochefoucauld.—19. Jarnac.—20. Lanau.—21. Belsunce.—22. Languedoc.—23. Noailles.—24. Schomberg.—All pockets were of the 'transverse' or horizontal type and bordered in the regimental colour (see the first illustration). At this time

The Mounted Chasseurs

The mounted chasseurs originated in Jean-Chrétien Fischer's[1] volunteers, who were known as 'Fischer's Chasseurs'. Formally organized as early as 1757, their squadrons were eventually attached to the dragoons from 1776, in the ratio of one squadron per regiment.

[1] Fischer was a German partisan leader in French service who distinguished himself in the War of Austrian Succession and was authorized to raise a company in 1743.

the cavalry regiments known by the names of Chartres, Condé, Bourbon, Conti, Penthièvre, Boufflers and Noailles were transformed into the 10th, 11th, 12th, 13th, 14th, 15th and 23rd Dragoons. The disbanded legions made up 24 squadrons of 'mounted chasseurs', and one of these was added as a fifth squadron to each of the 24 existing dragoon regiments. The livery braid decorating the horse's trappings in these converted cavalry regiments can be seen in the next plate, which refers to the year 1786, when dragoons lost their anomalous 'infantry' classification and came to be regarded as cavalry.

C. Dragoons and mounted chasseurs, statute of 1779. Regiments with white buttons had vertical pockets, those with yellow buttons had transverse ones. All pockets were decorated with piping in a contrasting colour except in the regiments numbered 25 to 30. The 1st to 6th Regiments of mounted chasseurs, however, had no pockets at all. The presence here of these intruders may come as a surprise, but it should be explained that they originated in the 24 squadrons attached to the dragoons in 1776 (see B above), which were later used to form six mounted chasseur regiments. These, which had no distinctive titles except their numbers, 1 to 6, nevertheless continued to form a part of the dragoons. Apart from the missing pockets, their uniform is generally considered identical to that of the dragoons, but some old sources show the mounted chasseurs wearing the tricorn hat with a red 'pineapple' crest.—1. Colonel General.—2. *Mestre de camp général*.—3. Royal.—4. King's.—5. Queen's.—6. Dauphin's.—7. Monsieur's.—8. Artois.—9. Orléans.—10. Chartres.—11. Condé.—12. Bourbon.—13. Conti.—14. Penthièvre.—15. Boufflers.—16. Lorraine.—17. Custine.—18. La Rochefoucauld.—19. Jarnac.—20. Lanau.—21. Belsunce.—22. Languedoc.—23. Noailles.—24. Schomberg.—Mounted chasseurs: 25. 1st Regiment.—26. 2nd.—27. 3rd.—28. 4th.—29. 5th.—30. 6th.

In 1779 they were regrouped into six regiments, with their own organization and numbers. Five years later each regiment was allocated a battalion of infantry, the total strength of each now comprising 612 horse and 348 foot. Foot and horse were separated again by a statute of 1788, and the number of regiments was increased to twelve by the conversion of six regiments of dragoons. A thirteenth regiment was raised in 1792, followed by thirteen more between 1793 and 1795.

The Hussars

The word 'hussar' comes from the Hungarian *huszar*, derived from *husz*, 'twenty'. But the reasoning behind this etymology remains debatable. The most popular theory is that each village in Hungary was obliged to provide one horseman for every twenty households. Others maintain that the Hungarian cavalrymen were paid twenty

FRANCE, DRAGOONS (IV)
A. 1786: On the left is shown the livery braid used by each regiment; the arrow shows the background colour of the saddlecloth. Regiments with yellow buttons had transverse pockets (figs. 1–3), those with white buttons had vertical ones (fig. 5). On either side of fig. 3 are shown the decorations used on the coat facings, fleur-de-lis for the front, grenade for the back.—1. Colonel General. The 1st Company, called the *compagnie colonelle-générale*, had white helmet crests and were mounted on grey horses. All other dragoons wore neatly curled black horsehair crests—at least in theory, though many documents show it worn loose, dangling free or even plaited at the back of the neck!—2. *Mestre de camp général.*—3. Royal.—4. King's.—5. Queen's.—6. Dauphin's.—7. Monsieur's.—8. Artois.—9. Orléans.—10. Chartres.—11. Condé.—12. Bourbon.—13. Conti.—14. Penthièvre.—15. Boufflers.—16. Lorraine.—17. Montmorency.—18. La Rochefoucauld.—19. Zweibrücken (Deux Ponts).—20. Durfort.—21. Ségur.—22. Languedoc.—23. Noailles.—24. Schomberg.
B. 1791: The statute of 1 January 1791 suppressed the old regimental names, and regiments were henceforth known only by their numbers. Next, a provisional order brought in the wearing of a small tuft of black horsehair on the helmet crest and white metal buttons for all regiments. The regimental colours were reduced to three, each allocated to a group of six regiments.
First group (shown here): 1. 1st Regiment (formerly Royal).—2. 2nd Regiment (formerly Condé).—3. 3rd Regiment (formerly Bourbon).—4. 4th Regiment (formerly Conti).—5. 5th Regiment, (formerly Colonel General's).—6. 6th Regiment (formerly Queen's).
The second group had crimson as its distinguishing colour, and its uniforms showed the same characteristics as the first group, in the same order, so it is simply necessary to replace scarlet with crimson in each case.—7th (formerly Dauphin's).—8th (formerly Penthièvre).—9th (formerly Lorraine).—10th (formerly *Mestre de camp général's*).—11th (formerly Angoulême).—12th (formerly Artois).
The third group, with pink as its distinguishing colour, follows exactly the same rules.—13th (formerly Monsieur's).—14th (formerly Chartres).—15th (formerly Noailles).—16th (formerly Orléans).—17th (formerly Schomberg).—18th (formerly King's).
The reforms of 1791 did away with the livery braid on the horses' trappings and brought in a standard, 'classless' braid of white thread. There also appeared at this time the sheepskin shabraque decorated with a saw-tooth border in the distinctive colour.
C. 'Hungarian-style' sword-belt of 1786—the buckle had no pin and was fastened with a hook underneath. Swords used were the 1770 (generally), 1781 and 1784 models (see figs. a, b and c). The bayonet was the 1777 model—note that the bayonet sheath held the weapon diagonally, as shown in fig. D.—D. A previous variation, used around 1780.—E. Sword-belt in 1754.—F. Sword-belt in 1750. The wide-bladed bayonet was soon abandoned in favour of the infantry version.—G. Sword-belt in 1720.—Masterly studies by Christian Ariès and Jean Boudriot have recently been produced dealing with French swords and fire-arms respectively.
1. Dragoon in greatcoat, 1786. The waistcoat was made up from worn-out greatcoats. The uniform coat, unadorned and very similar to the greatcoat, was used with full equipment, on horseback. The forage cap was decorated in the regimental colour.—2. Dragoon of the Artois Regiment with hooded cloak, 1786. Before this time, the cloak had only a heavy turn-up collar.—3. Dragoon of the 17th Regiment in 1791.—4. Dragoon of Durfort's Regiment in 1790. Note the curled helmet crest, still widespread at this period.
5. Helmet, 1763.—6. 1780 helmet with the 1791 crest.—7. Early nineteenth-century helmet.—Note the striking process of evolution which turned the strictly 'classical' helmet into its 'modern' descendant which would survive almost unchanged until 1914.

sous a day, but a Hungarian historian[1] has cast doubt on this idea, pointing out (among other things) that wages (the *zsold*) were always paid in kind, never in cash, even in the eighteenth century. On the other hand the smallest fighting unit of the Hungarian cavalry was the *husz*, or squad of twenty men, and the suffix *-ar* indicated a member of one of these units.

Under Louis XIII

The first Hungarian exiles served as auxiliary cavalry in the French army during the Thirty Years' War. From 1636 they were commanded by Georges Esterhazy, but their military value was obviously not yet fully realized.

[1] Charles d'Eszlary, Professor at the University of Pecs (Hungary), writing in the review of the International Hussar Museum at Tarbes (No. 3, 1968).

FRANCE, LIGHT HORSE AND MOUNTED CHASSEURS

A. 1779, light horse: 1. 1st Regiment.—1a. Epaulet and lanyard for 1.—2. Trumpeter, 6th Regiment.—3. a–e, 2nd–6th Regiments; f. officer's epaulet and counter-epaulet.—The six regiments all had braid in the royal livery on their horses' trappings (white chain stitch on crimson background, shown in several previous illustrations).—Mounted chasseurs: 4. Chasseur of the Alpine Regiment (1st), 1786. Regulations at the time prescribed an identical uniform for the other five regiments, but with distinctive colours as follows: 2nd (Pyrénées), crimson; 3rd (Vosges), pale yellow; 4th (Cévennes), buff; 5th (Gévaudan), yellow-gold; 6th (Ardennes), white.—4a. Epaulet and shoulder-strap, 1st Regiment. Cross-hatching, fringe and border are all in the appropriate colour for each regiment. The new hat, the *chapeau à visière*, with its brim turned straight up to the same height on both sides, was curiously similar to the Prussian *Kasket* of the same period. It was provided with a round, flat crest in which white was mingled with the appropriate colour for each company—1st, red; 2nd, sky-blue; 3rd, pink; 4th, marigold. The projected designs for the saddlecloth were to have braid in the royal colours.—5. Officer of the 3rd (Vosges) Regiment in 1789.—6. Trumpeter of the Normandy Regiment in 1789.
B. Colour-schemes of the 1st to 12th Regiments of mounted chasseurs as laid down in 1788. a. 1st, Alsace.—b. 2nd, Évêchés.—c. 3rd, Flanders.—d. 4th, Franche-Comté.—e. 5th, Hainault.—f. 6th, Languedoc.—g. 7th, Picardy.—h. 8th, Guyenne.—i. 9th, Lorraine.—j. 10th, Brittany.—k. 11th, Normandy.—l. 12th Champagne.—m. Hat and coat-facings, 1788.
C. Royal decision of 1789: a. 1st Regiment, Alsace.—b. 2nd, Évêchés.—c. 3rd, Flanders.—d. 4th, Franche-Comté.—e. 5th, Hainault.—f. 6th, Languedoc.—g. 7th, Picardy.—h. 8th, Guyenne.—i. 9th, Lorraine.—j. 10th, Brittany.—k. 11th, Normandy.—l. 12th, Champagne.—m. Facings in 1789.—The regiments were paired off to form brigades, the regiments in each pair being differentiated only by the colour of their buttons. 1st Brigade: 1st and 12th Regiments; 2nd Brigade: 2nd and 7th Regiments; 3rd Brigade: 3rd and 8th Regiments; 4th Brigade: 4th and 10th Regiments; 5th Brigade, 5th and 9th Regiments; 6th Brigade, 6th and 11th Regiments. Hats had plumes in the same colour as the uniform facings. For epaulets and shoulder-straps, see caption 4a.
D. Provisional ordinance of 1 April 1791: a. 1st Regiment.—b. 2nd.—c. 3rd.—d. 4th.—e. 5th—f. 7th.—g. 8th.—h. 10th.—i. 11th.—The regiments not shown will be found in figures 3, 4 and 5 of the central picture. It should be noted that the 6th Regiment appeared as in fig. d, but with the distinctive colouring on the collar only.
1. Chasseur of the 1st Regiment in marching order, 1791. He wears a uniform coat converted into a dress-coat, one of a hundred examples of imaginative improvisation to be found at this period.—2. Chasseur of the 1st Regiment in full dress, 1791. The carbine, it will be remembered, was suspended from a shoulder belt which was attached by a sliding snap-link to a rod running along the left side of the weapon. The number on the saddle bag was a tolerated irregularity. The shabraque was also found in green cloth with a white border, yet another instance of rule-bending!—3. Officer of the 9th Regiment in full dress, 1791. The silver braid round the shabraque varied in width according to rank (see text).—4. Trumpeter of the 6th Regiment in full dress, 1791.—5. Corporal of the 12th Regiment in 1791. The details of his helmet and the back of his uniform were common to all regiments.

Under Louis XIV

The 3,000 expatriates suddenly found themselves reduced to the most acute poverty in 1656 when Louvois, Louis XIV's war minister, decided to cut off their pay; he could no longer tolerate 'fellows like that' in the King's service. The lucky ones managed to find employment with such dignitaries as were attracted by the exotic eccentricity of their dress.

In 1691, however, one of the more enterprising hussars reported to Marshal François-Henri of Luxembourg[2] and offered to lead a raiding party of twenty[3] men to attack the enemy in the rear and pillage his baggage-train. When this sortie was a success the King promptly ordered the establishment of as many hussar companies as

[2] See Vol. I, pp. 26, 47 and 64.
[3] The strength of a company had always been twenty.

the number of refugees would allow. The news spread rapidly beyond the frontiers of France; from 1692 onwards deserters from the Imperial hussar regiments flocked to Louis's service.

The first regiment of Royal Hussars was immediately raised at Strasbourg and was placed under the command of an allegedly Hungarian but actually German 'baron' by the name of Kronenberg or Corneberg. This gallant officer was soon expelled from France for assorted malpractices, most notably losing part of the regimental funds at the gaming table. In 1693 he was replaced by a Bavarian gentleman, a former colonel in the Württemberg army, named Jacques-André de Mortany, who had been in the French service for several months. The regiment fought throughout the rest of the war until the Treaty of Ryswick (1697), when it was disbanded; 150 of its 300 horsemen were transferred to the Royal German Regiment.

So much for the first hussar regiment. Often criticized, ill disciplined and with no natural loyalty to their *mestre de camp*, the hussars had nevertheless proved their worth. They resurfaced in 1701, in the shape of a detachment of veterans of Mortany's regiment who had served with the Royal German Regiment. Then, in 1705, there appeared another force strengthened by deserters from the Imperial armies, the Marquis de Verseilles's Regiment.

It was in 1701, again, that a small 'regiment' of 140 hussars was presented by the Elector of Bavaria to the Marquis de Saint-Géniès. In 1707 this force came under the command of 'Baron' de Rattzki. Rattzki (or Raschi, Rattky or Rattsky, according to taste) was really a Czech adventurer by the name of Ratky or Hradki; his 'ancestral castle' of Salamanza in Hungary never existed.

In 1706 twelve independent companies serving in Italy were formed into a regiment under the command of a former colonel in the Imperial army who had been taken prisoner in France, Paul Déak, sometimes spelt Poldéak. The numerous complaints made against this unbalanced officer caused him to be cashiered and replaced by a German officer, de Feltz or de Filtz, who was killed in Rousillon the following year. His successor was the Comte de Monteils, who in 1709 took himself and his regiment off into the service of Louis xiv's grandson, Philip v of Spain. And so, by the end of Louis xiv's reign, France fielded two regiments of hussars, those of Verseilles and Rattzki.

FRANCE, HUSSARS (I)
1. Royal Hussar in the early eighteenth century. The axe was used to remove heads, which were kept as trophies. It is as well to remember that this barbaric custom was still quite widespread during the Second World War in the Balkans, especially against the Greek and Albanian partisans.—2. Royal Hussar in cloak, early eighteenth century.—3. Saint-Géniès Hussar, 1701. At first, each aigrette on the bonnet represented one enemy head taken; eventually there was no room left for more, and, with the price of aigrettes rising, it became customary to make do with one.—4. Officer of the Royal Hussars, 1700. These hussars were widely and quite accurately known as 'Hungarian Hussars'. The medieval horseman's hammer fulfilled the function of the officer's baton. The silver ball on the chest was a mark of nobility.— 5. One of David's (formerly Esterhazy's) hussars in 1744 (see next page, A, fig. 5). The horse's trappings, in ochre, were the same shape as those in fig. 6, and with the same decoration. —6. Bercheny's Hussars, 1745. The front of the shabraque is here turned back to keep the thighs warm. The sword shown is the 1730 model.—7. Officer's pelisse.—8. Trooper's pelisse. Pelisse and dolman very early on adopted the characteristic appearance which they were to retain everywhere until the end of the nineteenth century, and in some armies up until the First World War.
9. Guidon of Poldéak's Regiment in 1706. It measured 109 × 100cm. Poldéak, or more accurately Paul Déak, deserted in the same year. He was one of that regrettable crew of adventurers who, between 1692 and 1716, succeeded one another in command of France's first hussars.—10. Guidon of Polleresky's Regiment, captured (together with the regiment's seven other guidons) by the Prussian hussars in 1758. It measured 100 × 83cm, and was made of golden-yellow silk covered with a tight chessboard pattern of squares. Purists may note that the sun motif, with its exaggerated human features, did in fact have 24 rays.
11. Sabre, 1752 model.—12. Sabre, 1767 model.—13. Sabre, 1776 model.

FRANCE, HUSSARS (II)

A: 1. Rattzky's (formerly Saint-Géniès's—see fig. 3, previous page), 1707.—2. Bercheny's, 1720. A blue pelisse trimmed with fur is also found instead of the wolf-skin one.—3. Rattzky's in 1735, later to become Lynden's (Claude d'Apremont, Comte de Lynden, see C, fig. 4). In 1720 Rattzky is shown the same as Bercheny's (fig. 2), but with red breeches, blue *scharawades*, and the same shabraque as in fig. 1.—3a. Rattzky's in 1740.—4. Bercheny's in 1735.—4a. Bercheny's in 1740. The gold fleur-de-lis appeared on the shabraque in 1739.—5. Esterhazy's in 1734, later to become David's in 1744 (see preceding page, fig. 5), and then Turpin's (from Comte Turpin de Crissé) in 1746 (see C, fig. 2).—5a. Esterhazy's in 1740.

B: 6. Beausobre's, raised in 1743.—7. Raugrave's, raised in 1743.—8. Polleresky's, raised in 1743.—9. Ferrary's (from Ferrary of Interiani), raised in 1745.

C: 1744–52: 1. Bercheny's. In 1747 the white felt mirliton replaced the cap, except in Bercheny's Regiment, which obtained the King's permission to keep their old headgear. In 1755 a black mirliton replaced the white, which had proved too conspicuous, but once again Bercheny's managed to preserve its individuality by switching to red.—2. Turpin's (formerly David's) Hussars, 1746.—3. Polleresky's.—4. Lynden's.—5. Beausobre's. The silver braid of the 1747 white mirliton seems to have been replaced by royal blue.—6. Raugrave's, which became the Liège Volunteers in 1756.—7. Ferrary's.—In 1756 Lynden's, Beausobre's and Ferrary's were disbanded and their squadrons assimilated into Bercheny's, Turpin's and Polleresky's Regiments. The ordinance of 1744 prescribed 5 fleurs-de-lis on the shabraques, but we do not know how they were arranged, if indeed they ever existed.

The ordinances of 26 October 1744 and 15 May 1752 have long been the favourite target of detractors of Lienhart and Humbert, who, like other older sources, show braid and trimmings on dolmans and pelisses in the regimental colours, as we have them. This colour scheme, attacked today as a flagrant error, is accompanied by a note from the accused authors, who describe their view in the following prudent terms: 'If one is to believe numerous documents offering serious guarantees of authenticity . . .' a declaration which follows their citation of the statute clearly stipulated (p. 136) '. . . five and a half ells of white thread braid for buttonholes, and eight ells of white thread braid, six lines (13.8mm) wide, for sleeve and pocket borders.' Were the authors of the time right to attach credence to these serious documents, which may be no more than reflections of the 'Prussomania' then current in France? So unconditional was this admiration that it led to the shameful, and unrebuked exaltation of Frederick II's lightning victory at Rossbach by his French flatterers, led by no less than the great Voltaire! Is it such a rash assumption, then, to think it probable that coloured braid would be worn in the style of the Prussian hussars, bearing in mind that the entire French cavalry was to be converted into an imitation of Frederick's cuirassiers by the outbreak of the Seven Years' War? (See Cavalry, plate II).

D. 1758: 1. Bercheny's.—2. Turpin's, to become Chamborant's in 1761.—3. Polleresky's.—4. Royal Nassau, raised in 1758 from the Nassau-Sarrebrucken volunteers.—Polleresky's Hussars, generally believed to have been disbanded in 1757, were not in fact so until 1758, as a result of excessive looting and outrages against France's German allies. The fact that they allowed all their guidons to be captured by the Prussian hussars on 23 February of the same year was unlikely to encourage the high command to take a lenient attitude. The regiment was disbanded on 5 May and incorporated into Bercheny's and Turpin's.

E. 1762: 1. Bercheny's.—2. Chamborant's, formerly Turpin's, which took the name of its new *mestre de camp* from 1761.—2a. Sabretache (Chamborant's), 1767.—3. Royal Nassau. In 1767 golden-yellow replaced orange as the regimental colour.—4. Esterhazy's. In 1764 the King authorized Count Valentin-Ladislas Esterhazy to raise a new regiment, Esterhazy's Hussars. A previous regiment of the same name had ceased to exist in 1744 (see A, figs. 5 and 5a). This regiment at first wore white breeches.

F. 1776: 1. Bercheny's. On the right, the white-braided mirliton of 1779.—1a. Sabretache, 1779.—2. Chamborant's.—3. Conflans's, raised with the help of Conflans's Volunteers.—4. Esterhazy's. On the right, the white-braided mirliton of 1779.—The hussar regiments are no longer numbered in the cavalry sequence, but from now on form a special subdivision on their own. Note the disappearance of the Royal Nassau Regiment, reformed in 1776.

G. 1786: With their new status as a special branch of the cavalry, the hussars were given a Colonel General's Regiment, raised in 1783.—1. Colonel General's.—2. Bercheny's.—3. Chamborant's.—4. Conflans's. This regiment became the Saxony Hussars in 1789, and emigrated almost to a man in 1792.—5. Esterhazy's.—6. Lauzun's, formed from Lauzun's foreign volunteers, who fought with the American rebels and became Lauzun's Hussars in 1783.—In 1791 the hussar regiments lost all distinctive titles except the numbers showing the order in which they were raised; thus Bercheny's became the 1st, Chamborant's the 2nd, Esterhazy's the 3rd, the Saxony Hussars the 4th, the Colonel-General's the 5th and Lauzun's the 6th.

H. Sabretache of Bercheny's *mestre de camp*, 1764.—I. Sabretache of a senior officer, Colonel General's Regiment, 1786.—J. Sabretache of a junior officer, Conflans's Regiment, 1788.

Under Louis XV

During the minority of Louis xv and the regency of Philip of Orléans, in 1716, Verseilles's regiment was amalgamated with Rattzki's.

In 1720 Ignace-Stanislas de Bercheny was authorized to raise a new regiment from the hussars who had ridden with him from Constantinople the year before. Bercheny's career in the French cavalry was to last nearly sixty years; one of the great men of the service, he eventually became Lieutenant General of the King's Armies and Inspector General of Light Cavalry.

In 1734 another hussar regiment was raised at Strasbourg, by Count Ladislas-Ignace Esterhazy. But subsequent financial difficulties forced him to hand over command to his second-in-command, M. David, also a native of Hungary. Twenty years later another genuinely Hungarian regiment appeared under the command of Valentin-Ladislas d'Esterhazy.

In the meantime, Polleresky's Regiment had a brief existence: raised in 1757, it was disbanded in 1758 for looting and robbery on service in Germany. These depredations had become rare since the rise of Bercheny, whose own regiment was a model of discipline. Before his time, the hussars had frequently been subject to severe criticism. Marshal de Lacolonie said of them in 1703, 'They are nothing more nor less than mounted bandits.' In 1711 Gassion wrote: 'How many standards were captured I cannot say, as they were taken by the hussars, who melt down the gilding for their own purposes.' Their officers had long been obliged to give them plenty of rope; the alternative was, quite simply, to be murdered. No officer would have dared to order a charge without first obtaining the approval of the men; an over-confident officer would have found himself charging alone!

Hungarian recruits being in short supply, the regiments were kept up to strength with Poles and Turks, later with Germans and even German-speaking Frenchmen from the Rhine provinces. Officers had to know German, as orders were always given in that tongue, though all learnt to swear convincingly in Magyar. By the time the monarchy fell, discipline in the hussar regiments had become very harsh, with a severity unparalleled in any of the other cavalry regiments.

FRANCE, HUSSARS (III)
1. Trumpeter of the Royal Nassau Regiment, 1767.—2. Trumpeter, Chamborant's Regiment, 1784.—3. Trumpeter, Bercheny's Regiment, 1760.—4. Trumpeter, Colonel General's Regiment, 1786. The embroidery in the colours of the Duc d'Orléans, the future Philippe-Égalité and Duc de Chartres until 1785, was worn only on the sleeves and the 'swallows' nests' on the shoulders. The rest of the pelisse was trimmed with blue, white and red. The Orléans embroidery will be found in the cavalry illustrations.—5. Trumpeter, Lauzun's Regiment, 1786.—All these trumpeters, apart from No. 4, had pseudo-sleeves hanging loose behind them, in the fashion so popular during the eighteenth century.—6. Foot chasseur attached to Bercheny's Regiment, 1760. Those attached to Turpin's Regiment had the same uniform but with black cuffs and facings.
The mirliton: 7. Parade dress, with the *flamme* hanging free. —8. Full dress. The *flamme* is rolled up, with the lining outwards.—9. Campaign dress, the *flamme* again being rolled up, but now with the black side outwards. Veteran hussars would sometimes fashion themselves a sort of peak by rolling the *flamme* less tightly.—10. Hussar of Bercheny's Regiment wearing a variation of the mirliton. The sliding rings on the sash were sometimes arranged in diagonal lines, or chess-board-fashion.
11. Officer of Chamborant's Hussars with undress hat and pelisse worn without the dolman, as fashion demanded around 1779.—11a. Regulation shabraque; the animal's head, minus its lower jaw, always covered the horse's cruppers. In 1789 the shabraque had a trimming of gold or silver braid, its width varying according to rank, sewn around the edge of the fur, inside the scalloped border. As can be seen, hussar officers of this period were a match for their earliest predecessors in splendour, if not in barbarity.—11b. Sabretache of a senior officer in Bercheny's Regiment around 1790.—12. Assistant Quartermaster Sergeant, Conflans's Regiment, 1786. As well as the gold braid (silver for those regiments which wore white buttons) on the sleeves, the pelisse was bordered with fox-fur, a custom shared also by C.S.M.s and Sergeants.—12a. Sergeant Major.—12b. Sergeant.—12c. Corporal.—12d. Lance Corporal.—13. Colonel General's Regiment, with the plume adopted in 1791. The mirliton was attached to the body by a cord passed round the chest or neck (see adjacent illustration).—13a. Campaign stripe, worn on the left arm; the maximum number that could be worn was two, and their colour matched that of the braid.— 14. Adjutant, Colonel General's Regiment, 1786. In 1791 the braid surrounding the frogging, a distinction peculiar to this regiment, disappeared.—14a. Arrangement of the adjutant's gold chevrons on the pelisse. Left: mirliton, showing the cord and the way in which it was attached.
Trumpeters' livery embroidery: A. Bercheny's.—B. Chamborant's.—C. Conflans's.—D. Esterhazy's.—E. Lauzun's.

Under Louis XVI

The establishment of the Colonel General's Regiment is often represented as a favour conferred by the King on his cousin, Louis-Philippe-Joseph d'Orléans, Duc de Chartres.[1] In fact it was the reverse, a ruse to deny him the naval career he hoped for. The regiment was officially founded in 1779, but did not take the field until 1783, while the Duke's contribution was confined to designing a uniform which was never worn; the actual command he left to his senior *mestre de camp*. In the ranks of this regiment there served two men whose names were to become household words under Napoleon's Empire: Kellermann and Ney.

The Wallace Collection, in London, has a superb portrait of the Duc de Chartres in the uniform of Colonel General of Hussars, painted by Reynolds.

---

[1] He was to become Duc d'Orléans on the death of his father in 1785. Known as Philippe-Égalité, he voted for the death of Louis XVI; his son became King of France as Louis-Philippe.

FRANCE, LIGHT TROOPS

1. Fusilier of the Breton Volunteers, raised by the Chevalier de Kermellec-Penhöet in 1746. In 1749 this regiment was incorporated into the Flanders Volunteers. The 'hussars' attached to this unit wore the same uniform with a blue shabraque with light yellow border. — 2. Soldier of the Cravate Infantry, raised by Marshal de Saxe in 1746. This regiment was demobilized in 1748 after the war of Austrian Succession. It had been formed from Austrian deserters. — 3. Dragoon and Fusilier of de la Morlière's Regiment. Raised in 1745 this force distinguished itself in the Flanders campaign and became known as the Flanders Volunteers. The dragoons had shabraques of white sheepskin bordered with a yellow saw-tooth pattern. — 4. Uhlan of Saxe's Volunteers, raised in 1743. The sheepskin shabraque was bordered with red. — 5. Schomberg's Volunteers, formed from veterans of Saxe's Dragoons in 1750. Their conversion to the 17th Dragoons in 1762 gave the dragoons their famous 'Schomberg helmets'. The dark green shabraque had an orange border and a white fleur-de-lis in the corner. — 7. Officer of Royal Volunteers. The saddlecloth was red with a white border and a white fleur-de-lys in the corner. — 7. Officer of Fischer's Chasseurs in 1745. This first French force to use the name 'chasseur' was raised in 1743 by a former servant who had rebelled against the Austrians, and made himself the leader of a small fighting force. Having no coat of arms, Jean-Chrétien Fischer chose for his emblem three fish, which fitted his name well enough. — 8. Dragoon of Marshal de Saxe's Volunteers, 1745. This cavalry regiment, raised in 1743, was the nucleus from which Schomberg's Dragoons were later formed (1755), and included as many dragoons or *pacholeks* as it did uhlans. The *pacholek* was originally the valet of an uhlan in the Polish *pulks* on which this force modelled itself. — 9. Mounted and unmounted companies of Grassin's Arquebusiers in 1745. This infantry force, raised in 1744, was commanded by Simon-Claude de Glatigny de Grassin and distinguished itself particularly at the Battle of Fontenoy. The cavalry companies appeared in 1745. In 1749 the Arquebusiers became part of the Flanders Volunteers. — 10. Uhlan of Marshal de Saxe's Volunteers, 1745. The uhlans were all of noble birth, though it may seem difficult to accept this when one recalls that Maurice de Saxe's personal bodyguard was formed of superbly uniformed Negroes. They aroused the criticism of envious individuals who revolted against this dangerous intrusion of 'armed slaves in the very heart of the French forces'. The uhlans did guard duty at the chateau of Chambord, Marshal de Saxe's residence. Uhlans and dragoons wore crests in the distinctive colours of their respective brigades, which could be white, red, yellow, blue, green or black.

11. Foot-soldier and cavalryman of Clermont-Prince's Foreign Volunteers, raised in 1758; they took the name of 'Condé's Legion' in 1766. Their horses wore the classical dark blue saddlecloth. — 12. Condé's Legion in 1766. The short boots are typical of the period. The shabraque was in sheepskin, bordered with buff cloth. — 13. Member of the Perpignan bourgeois militia in 1758. Note the cap called the 'baratina' and the hairnet. — 14. Mountain Fusiliers in 1745 and 1758. This regiment was raised in 1745 and disbanded in 1762. Note the weapons used by these ancestors of the Alpine infantry, which were their own. — 15. Cantabrian Volunteers, established in 1745.

The Light Troops

At the beginning of the eighteenth century the only light troops in the French army were the mountain fusiliers. The years from 1727 onwards however, saw the raising of several independent companies. Their value as skirmishers became so rapidly obvious that their numbers grew continuously, especially after the outbreak of the War of Polish Succession in 1733 and that of Austrian Succession in 1740. The following were the main forces to appear during this period:

- 1743 Saxon Dragoons, Fischer's Chasseurs[1]
- 1744 Grassin's Arquebusiers
- 1745 La Morlière's Fusiliers, Cantabrian Volunteers; a number of independent companies combined to form a single corps of Royal Volunteers
- 1746 Gantès's Volunteers, Breton Volunteers, Croatian Infantry, Dauphiné Volunteers
- 1747 Hainault Volunteers
- 1749 Flanders Volunteers (formed from Grassin's Arquebusiers, Breton Volunteers and la Morlière's Fusiliers)
- 1756 Royal Legion (former Royal Volunteers)
- 1757 Flanders Volunteers and Hainault Volunteers (i.e., the Flanders Volunteers were split into two regiments)
- 1758 Clermont-Prince's Volunteers
- 1761 Conflans's Dragoon Chasseurs (former Fischer's Chasseurs)
- 1762 Flanders Legion (Flanders Volunteers and Dauphiné Volunteers), Hainault Legion (Hainault Volunteers)
- 1763 Royal Legion (Royal Volunteers); Conflans's Legion (Conflans's Dragoon Chasseurs)

[1] See above, in the section on the mounted chasseurs.
[2] See Vol. I, p. 24.
[3] A village in the Halle district of eastern Germany where in 1757, during the Seven Years' War, Frederick II beat the French and their German auxiliaries.

In 1776 the Comte de Saint-Germain disbanded the light troops, and their cavalry was transferred to the dragoon regiments as chasseur squadrons.

Despite the chocolate-box soldier appearance of most of these units their members fought magnificently. The infantry was composed mainly of young, fit recruits, while the cavalry was composed of battle-hardened veterans who never deserted their infantry comrades. In the triumph of Fontenoy[2] or the shame of Rossbach,[3] the light troops never failed to acquit themselves honorably. Their effectiveness, so often demonstrated, considerably slowed the development of properly organized light infantry and cavalry regiments. These did not come on the scene until the days of the Republic, which they served with a style worthy of their predecessors.

FRANCE, ARTILLERY
1. Engineer, 1745.—2. Officer of the Royal Artillery Corps, 1745.—3. Artillery officer, 1745.—4. Artillery train, 1745.—5. Artillery labourer, 1757.—6. Artilleryman, 1757. Breeches were red, gaiters black or white according to circumstances and to the time of year.—7. Geographical engineer, 1780.—8. Trooper of the Royal Artillery Corps, 1772.—9. Royal Artillery Corps, 1786.—10. Artilleryman, 1792. The sapper had the same uniform but with black collar, cuffs and facings.—11. Sapper, 1757.—12. Sapper, 1786.—13. Officer commanding a labour company, 1786.—14. Labourer, Royal Artillery Company in 1786 (working dress).—14a. Labourer's coat (Royal Artillery Company, 1786).—15. Horse Artillery, 1792.

The Artillery

In France, as everywhere else during the early eighteenth century, the artillery was dependent on very cumbersome cannon of varying sizes – 4, 8, 12, 16, 24 and 36 pounders – and 6, 7, 8, 9, 10 or 11 pound mortars. The strength of the Artillery Corps was at that time 697 officers and 5,630 men.

Not until 1732 did Lieutenant General de Vallière,[1] the first Inspector General of Artillery, see his own rationalized system brought into effect: only 4, 9, 12, 16 and 24 pound cannon were to be constructed in future, the first three being suitable for use as field artillery while the complete range would be adequate to defend or attack a fortified position; the field artillery assisting the defence of the citadel, and the citadel in turn lending some of its own cannon to the army, if needed, for use in the field.

Mortars were now made in two sizes, 8-inch and 12-inch, and a 15-inch bombard was also introduced. However despite these efforts with the shape, size and proportions of the guns, Vallière had done nothing to improve the gun-carriages, which were extremely heavy and varied from one part of the country to another.

In 1757 a light four-pounder, known as a 'Swedish cannon', was issued to the infantry, one per battalion. This was a very sensible innovation, but it needed all the authority of Marshal de Saxe[2] to overcome the resistance of the artillery-men. This gun weighed only 150 times the weight of its own shot (compared to 280 times for the old four-pounder), and was particularly useful during the Seven Years' War, when the rest of the artillery had available only whatever cannon could be constructed on an ad hoc basis.

The much-needed reform was introduced in 1765 by Jean-Baptiste Vaquette de Gribeauval, recalled by Choiseul from the Austrian service. This brilliant officer divided the artillery into four classes: field, garrison, siege and coastal. In the teeth of violent opposition Gribeauval 'specialized' his cannon, reducing their length and weight without impairing their range. The charge, which had a fixed ratio of one-third the weight of the projectile, was incorporated into it by the invention of cartridge, which tripled the rate of firing, while the introduction of adjustable sights led to much improved accuracy. Turning next to the gun-carriages, Gribeauval standardized them and made them lighter in the process, by separating the carriage from the limber he made it infinitely more easily manoeuvrable. An efficient repair service, spare parts and a hundred other improvements made the French artillery an incomparably efficient tool. Artillery schools were opened at la Fère, Metz, Strasbourg, Valence, Douai and Auxonne.

The fall of Choiseul in 1772 gave the temporary ascendancy to Gribeauval's enemies, led by the son of Vallière.[3] The 'reds' of the old school and the 'blues' of the new clashed head on in this so-called 'artillery duel', but the advent of Saint-Germain gave victory to the 'blues' in 1784. The great Gribeauval continued working on his guns until his death in 1789. As for the guns themselves, they gave faithful service until as late as 1825!

The Horse Artillery

In 1791 the Constituent Assembly voted to set up 'flying batteries' at the instigation of La Fayette,[4] who had seen them used on manoeuvres in Silesia. Two horse batteries were set up in 1792, followed by seven more a few months later. After the battle of Jemappes[5] in November of the same year there were thirty of them.

---

[1] Jean-Florent de Vallière (1667–1759).
[2] Maurice de Saxe, Marshal of France (1696–1750), the victor of Fontenoy. The illegitimate son of Augustus II of Saxony (see Chapter on 'Saxony', p. 82) he was to have a famous great-grand-daughter in a very different field, George Sand. See also the chapter 'The Walloon Regiments', p. 98.

[3] It was at this time also that the 1770 Gribeauval rifle was abandoned in favour of the 1773 Vallière model.
[4] Marie-Joseph Morier, Marquis de La Fayette (1757–1834). Hero of the American War of Independence who played a political role in the early days of the French Revolution.
[5] Or Jemmapes, near Mons, where the French under Dumouriez defeated the Austrians, a victory which resulted in the annexation of Belgium to France.

BRITISH CAVALRY AND ARTILLERY

The third part of Volume I was devoted to a review of the British infantry. We now return to Britain, for a look at the dazzling array of cavalry which, like the British infantry, distinguished itself in the many wars of the eighteenth century. And, as with France, we shall include a chapter on the artillery.

The Life Guards

It was during the exile of Prince Charles – later Charles II – that the Life Guards were formed, by the efforts of eighty Royalist gentlemen. On his return from France in the following year Charles brought with him three companies of guards, amounting to 600 men in all: the King's Troop, the Duke of York's Troop (the Duke was of course Charles's brother, the future James II), and the Lord General's Troop.[1] In 1661 these companies became known as, respectively, 'His Majestie's own Troope of Guards', 'His Highness Royall the Duke of Yorke his Troope of Guards', and 'His Grace the Duke of Albemarle his Troope of Guards' – the Lord General, Monck, had now become the Duke of Albemarle. On his death in 1670 his troop became known as the Queen's Troop.

A fourth company of Life Guards was raised in Scotland at the time of the Restoration, in 1660, but did not join the other three in London until 1707, following the Act of Union between England and Scotland.

In 1746 the Life Guards were reduced to two companies; later, in 1788, they were reconstituted as two regiments, the 1st and 2nd Life Guards.

The Horse Grenadiers

In 1678 a division of mounted grenadiers was added to the strength of each of the three companies of Life Guards stationed in London. These horse grenadiers were in effect dragoons – they looked and fought very like dragoons, but with the grenade as an extra weapon. They were certainly of inferior status to the Life Guards with whom they served, their officers being paid only half what Life Guards officers received.

After being disbanded briefly in 1680 the Horse Grenadiers reappeared in 1683, and two companies survived until the Life Guards were reorganized in 1788; they were then dispersed among the line cavalry regiments.

The 'Heavies'[2] – Royal Horse Guards

This regiment, now part of the Household Cavalry, was not officially declared a Guards regiment until 1820, as a result of its outstanding conduct on the field of Waterloo. It had been formed in 1660, when the soldiers who had served under Unton Crook, one of Cromwell's colonels, were reorganized under the Earl of Oxford.

---

[1] The Lord General was the famous Monck – see Vol. I, p. 122.
[2] As opposed to light cavalry such as the Dragoons.

Christened 'Royal Regiment of Horse', the new regiment was immediately nicknamed the Blues, after the uniform it always wore. From 1690 this became 'the Oxford Blues' or 'Lord Oxford's Blues' to distinguish them from William III's imported Dutch Horse Guards, who also wore blue. In 1687 their official title was changed to Royal Regiment of Horse Guards, but although they were very often used for the same escort duties as the Life Guards, they were nevertheless not part of the Guards; they were simply considered as the senior of all the 'Horse' regiments.[1] This seniority is clearly recognized in the famous *Clothing Book* of 1742,[2] which refers to the Royal Horse Guards as 'the 1st Regiment of Horse'. This description, though no doubt intended as an honour, seems not to have found favour. In 1746, when various regiments of horse were converted to Dragoon Guards, the 5th Regiment of Cavalry was given the number 1.[3]

The Royal Horse Guards thus enjoyed the signal honour of an unnumbered position between the Household Cavalry and the new Dragoon Guards. This ambiguous situation persisted until 1820; and, strangely, the regiment is not mentioned in any 'cavalry list' throughout the rest of the eighteenth century. Evidently Charles II's years of exile in France gave him the idea of making the Life Guards and the Royal Horse Guards into English equivalents of two crack French units, the *gardes du corps* and the *gendarmerie* of the Royal Household.[4]

In 1760 the Royal Horse Guards adopted a custom which set them apart from all other regiments, that of saluting even when bare-headed. The Marquis of Granby, Colonel of the Blues, had contrived to lose both hat and wig in the course of a triumphant cavalry charge at Warburg,[5] hence the origin of this odd tradition.

GREAT BRITAIN, ROYAL HOUSEHOLD (I)

1. 1st Company of the Life Guards in 1742. The men of the Life Guards, regarded as 'gentlemen troopers', wore their hair long and powdered, as did their officers. The four companies were distinguishable by the colour of the braid crossing their carbine belts, which colour was also to be found in the horse's trappings: 2nd company, white; 3rd, yellow; 4th, blue. — 2. 1st Company of the Life Guards in 1751. The Life Guards were reduced to two companies in 1746. The 2nd Company had two blue bands on its shoulder-belts; saddlecloth and holster-covers were identical to those of the 1st Company. Buttonholes were theoretically grouped in pairs in both units, but are shown evenly spaced in a contemporary picture of a trooper of the 1st Company; also, it seems that the gold braid trimmings had a red central line in the 1st Company and a blue one in the 2nd. — 3. 1st Company, Horse Grenadiers, in 1742. Recruited from the common people, the Grenadiers wore their hair like the line cavalry, i.e. gathered up under the hat. The carbine, carried as shown in fig. 1, was a weapon derived from the infantry's classic 'Brown Bess', and used the same standard flintlock. This type of weapon is now extremely rare. The 2nd Company can most easily be distinguished from the 1st by replacing the blue of the horse's trappings with red. — 3a. Cartridge pouch, 2nd Company of the Horse Grenadiers in 1750. — 4. 1st Company of the Horse Grenadiers in 1750. The carbine is carried here as for the parade-ground instruction, 'Advance your muskets.' The lanyard holding the powder flask was red in the case of the 2nd Company, whose horses also had red trappings with a yellow border marked with a blue line. The outer edge was also bordered with a thin red braid. — 4a. Variations in cuffs. — 5. Mitre cap of the 1st Company, 1750. — 6. Mitre cap of the 2nd Company, 1750. — 7. Mitre cap of the 2nd Company, 1742. — 8. 1st Company, Life Guards of Horse, in 1712. — 9. Variation in cuffs (the right cuff is shown here) in the 2nd Company, 1751. The Life Guards' coat pockets were horizontal, and had a bracket-shaped flap trimmed with gold braid. — 10. Exact reproduction of the collar of fig. 1, from the 1742 *Clothing Book*. No one has ever been able to determine the precise nature of this piece of cloth. Our own opinion is that it is simply the blue lining of the coat as shown in our fig. 1. Note the overlapping lapels in our fig. 2, which seem to support this theory. — 11. R. Money Barnes's interpretation, from his excellent *Soldiers of London* (Seeley Service & Co., London, 1963).

---

[1] The 'Horse' regiments were simply the heavy line cavalry. The exceptional status of the Royal Horse Guards meant that they were neither a Horse regiment nor a part of the Household Cavalry.

[2] See Vol. I, p. 104.

[3] See tables in next chapter 'The Regiments of Horse', p. 46.

[4] According to Major R. Money Barnes in *The British Army of 1914*.

[5] Near Minden in western Germany. One of the battles of the Seven Years' War, where Duke Ferdinand of Brunswick defeated the French in 1760.

The Regiments of Horse

The first regiments of horse (cavalry) were raised in 1685, in the reign of James II, to suppress the rebellion led by Charles II's illegitimate son, the Duke of Monmouth.[1] Their names then were:

'The Queen's' or '2nd Regiment of Horse'
'The Earl of Peterborough's Regiment of Horse' ('3rd' from 1688)
'The Earl of Plymouth's Regiment of Horse' ('4th' from 1687)
'The 6th Horse' or 'The Earl of Arran's Cuirassiers' ('5th' from 1690)
'The Duke of Shrewsbury's Regiment of Horse' ('6th' from 1687)

By 1742 the regiments were:
'The Royal Horse Guards'
'The King's Own Regiment of Horse'
'The Queen's Regiment of Horse'
'4th Regiment of Horse'
'5th Regiment of Horse'
'6th Regiment of Horse'
'The King's Carabineers' (raised in 1688)
'8th Regiment of Horse' (raised in 1688)

In 1746 the King's (2nd), Queen's (3rd) and 4th Regiments became respectively the 1st, 2nd and 3rd Dragoon Guards (see next chapter), while the other four regiments remained 'regiments of horse', as follows:

'1st Horse' ('5th' in 1742)
'2nd Horse' ('6th' in 1742)
'3rd Horse' or 'The Carabineers' ('7th' in 1742)[2]
'4th Horse' ('8th' in 1742)

In 1768 these last survivors of the old Horse Regiments became in turn Dragoon Guards, taking the regimental numbers 4, 5, 6 and 7.

---

[1] See Vol. I, p. 124.
[2] Soon rechristened the Irish Horse – a simple economical device to reduce their pay, which was fifty per cent lower for troops stationed in Ireland!
[3] See Vol. I, p. 114.

The Dragoon Guards

As we have just seen, the origin of the Dragoon Guards was the conversion of the first three cavalry regiments in 1746. This may seem surprising, but the purpose of the change was more one of simple economy than of developing a new type of soldier; a dragoon received only half the pay of a cavalryman. This step was no doubt justified by the expenses incurred in the suppression of the Jacobite rising of 1745.[3] The victims of the economy were compensated by being able to regard themselves as 'Guards', which did nothing for their pockets but at least preserved their seniority. Apart from some differences in uniform, the new dragoons differed from the horse regiments in their more frequent use of the musket. By the time the surviving four regiments of horse were added to the Dragoon Guards in 1788, a dragoon's pay had fallen to 1s. 6d. per day ... but the dragoons might feel themselves well off compared to the infantry, who were paid a mere $6\frac{3}{4}d$.

GREAT BRITAIN, ROYAL HOUSEHOLD (II)
1. Horse Grenadier Guard in 1787. — 2. Officer of the Horse Grenadiers in 1787. The sword-belt was subsequently worn crosswide, not passing underneath the crimson cloth sash. — 3. 2nd Company of the Horse Grenadier Guards in 1787. The interior of the fur cap may possibly have been blue in the 1st Company. — 4. Negro trumpeter of the Life Guards in 1750. Trumpeters were considered as non-combatants, and equipped only with broken swords. — 5. Life Guards in 1768. — 6. Officer of the Life Guards in 1797. — 7. Officer of the Life Guards in 1797. The common soldier's uniform was similar but without the crimson silk sash and the epaulets. The 1st Regiment had a blue collar, the 2nd a red collar with a blue piece at the front. The abandoning of the emblem of the Garter for the Star of the same order coincides with the reorganization of the Life Guards on 26 March 1788. — 8. Officer of the Life Guards, 1756. — 9. Officer of the Life Guards in 1768. The coat lapels are here unfastened and turned inwards to button the coat across the chest. — 10. Officer of the Life Guards in 1770. — 11. Officer of the Life Guards around 1800, with the characteristic crescent-shaped bicorn hat.

The Dragoons

James II had had dragoon regiments in his army since 1685. At one time there were no less than fourteen of them, but this figure began to dwindle after the end of the Seven Years' War. One after another, they were converted to light dragoons until only the six oldest regiments survived; in 1799 they were further reduced to five, when the 5th Irish Dragoons were disbanded following their involvement in the Irish Rebellion. These regiments, and the names by which they were known, are given in the table on page 50.[1]

The famous 5th Regiment, the Royal Irish Dragoons, was the first of two dragoon regiments raised at Inniskilling[2] during the Irish War; for this reason the second of the two, the 6th, was given the name of 'Inniskilling Dragoons' in 1751. The 5th Regiment later fought on the Continent, at Blenheim, Oudenaarde, Malplaquet, Fontenoy[3] and elsewhere, but the Irish troubles of 1798 were to be its death warrant. The regiment, aided by loyalist troops, had taken the town of New Ross when it was forced to recruit local men to make up the losses sustained in the heavy fighting. A number of dedicated rebels managed to enlist without arousing the suspicions of the recruiting officers, with the specific aim of massacring the officers and various 'traitors' within the regiment. Fortunately the plot was discovered in time, but the severe fright it caused was enough, and on 8 April 1799 the regiment was disbanded at Chatham.[4]

Originally mounted infantry, the British dragoons soon became – like their counterparts everywhere else – a sort of medium cavalry, without the cuirass worn by the 'heavies' and, just as important, very much cheaper to maintain. They soon abandoned the characteristic dragoon fighting tactics in favour of traditional cavalry methods. The famous Greys, the 2nd Dragoons, fought on foot in the Danube campaign in 1704, but did not repeat the experience until 1914! This departure from their original function became more and more marked after the beginning of the eighteenth century. A few regiments, however, produced peculiar grenadier companies, ostensibly to keep alive the dragoons' infantry traditions, but in fact more to enable them to wear the coveted grenadier mitre. This flight of fancy was short lived except in the Scots Greys (2nd) and Royal Irish (5th) regiments – and even here we have been unable to discover any pictorial evidence of it, whatsoever.

Although the dragoons virtually always fought as cavalry they continued to carry the bayonet: not until 1796 were they sufficiently convinced of the drawbacks of the cumbersome, ineffectual infantry rifle to abandon it in favour of the cavalry carbine.

GREAT BRITAIN, LINE CAVALRY (I)
A. 1742: 1. 1st Regiment or Royal Horse Guards. — 2. 2nd Regiment or King's Horse. — 3. 3rd Regiment or Queen's Horse. — 4. 4th Regiment. — 5. 5th Regiment. W. Carman gives the saddlecloth and cloak as blue, the latter lined with red. — 6. 6th Regiment. Lawson shows the background of the saddlecloth as white. 7. 7th Regiment, or King's Regiment of Carabineers. — 8. 8th Regiment, known as the 'Black Horse'.
B. 1751: 1. Royal Horse Guards. In 1746 this regiment lost its number, while the 2nd, 3rd and 4th Regiments became the 1st, 2nd and 3rd Dragoon Guards. — 2. 1st (Kings) Dragoon Guards (formerly King's Horse, No. 2 above). — 3. 2nd (Queen's Regiment) Dragoon Guards (formerly 3rd Horse). — 4. 3rd Dragoon Guards (formerly 4th Horse). — 5. 1st (formerly 5th) Horse. — 6. 2nd (formerly 6th) Horse. — 7. 3rd (formerly 7th) Horse. — 8. 4th (formerly 8th) Horse. As will have been noted, the conversion of three cavalry regiments into Dragoon Guards entailed the renumbering of the other cavalry regiments. — Note that in the case of the Dragoon Guards regiments the coat facings stop at the waist, whereas those of the line cavalry continue down to the bottom of the coat. The 2nd Dragoon Guards is alone in having its buttonholes set in threes.

---

[1] Based on R. Money Barnes, *The British Army of 1914*.
[2] Or Enniskillen, a town in Northern Ireland which resisted James II's troops in 1689.
[3] See Vol. I, pp. 24, 26 and 94.
[4] These previously unpublished details we learnt from our good friend Glenn Thomson, artist-historian of the Irish army. New Ross is a town in south-east Ireland; Chatham, in Kent, was Britain's leading naval base under Charles I and Charles II.

DATE RAISED	FIRST NAME	LATER NAMES	NICKNAME
1661	Tangier Horse	1683: King's Own Royal Regiment of Dragoons 1690: Royal Regiment of Dragoons 1751: 1st Royal Dragoons	The Tangier Cuirassiers
1681	Royal Regiment of Scotch Dragoons	1707: Royal North British Dragoons 1751: 2nd or Royal North British Dragoons	The Greys Dragoons[1]
1685	Queen Consort's Own Regiment of Dragoons	1692: Queen's Dragoons 1714: King's Own Regiment of Dragoons 1751: 3rd (King's Own) Dragoons	Bland's Dragoons[2]
1685	Princess Anne of Denmark's Dragoons	1751: 4th Dragoons 1788: 4th or Queen's Own Light Dragoons	
1689	Wynne's Dragoons	1704: Royal Dragoons of Ireland 1751: 5th Royal Irish Dragoons 1799: Disbanded – number vacant till 1858	
1689	Cunningham's Dragoons	1751: 6th or Inniskilling Dragoons	
1689	Queen's Own or Robert Cunningham's Dragoons	1751: Princess of Wales's Own Light Dragoons	Strawboots
1693	Henry Cunningham Dragoons	1751: 8th Dragoons 1775: 8th Light Dragoons 1777: 8th or King's Royal Irish Light Dragoons	The Cross Belts[3]
1751	Wynne's Dragoons	1751: 9th Dragoons 1783: 9th Light Dragoons	
1715	Gore's Dragoons	1751: 10th Dragoons 1783: 10th or Prince of Wales's Own Light Dragoons	
1715	Honeywood's Dragoons	1751: 11th Dragoons 1783: 11th Light Dragoons	
1715	Bowles's Dragoons	1751: 12th Dragoons 1768: 12th or Prince of Wales's Light Dragoons	
1715	Munden's Dragoons	1751: 13th Dragoons 1783: 13th Light Dragoons	
1715	Dormer's Dragoons	1720: 14th Dragoons 1776: 14th Light Dragoons 1798: 14th or Duchess of York's Own Light Dragoons	

[1] The name 'Scots Greys' did not become official until 1877.
[2] Bland was the colonel commanding the regiment at Dettingen in 1743. See Vol. I, p. 104.
[3] See fig. 8, p. 55.

GREAT BRITAIN, LINE CAVALRY (II)

1. Royal Horse Guards in 1760 (showing the cartridge-pouch, which was worn at the front).—2. Royal Horse Guards in 1793.—3. Royal Horse Guards in 1796. On their return from service in Europe they lost their saddlecloths and holster-covers, except for ceremonial and parade-ground dress. The horses had had their tails left uncropped after 1764, except in the case of the Light Dragoons.
Dragoon Guards: 1st Regiment in 1764 (4), in 1765 (5), in 1768 (6) and in 1774 (7).—2nd Regiment in 1768 (8) and in 1783 (9). The horses of the 2nd Dragoon Guards were bays (i.e. brown with black manes and tails) from 1762 onwards, whence the name of 'Bays' or 'Queen's Bays' by which the regiment was known at this time.—10. 3rd Regiment in 1768.—11. 3rd Regiment in 1788.—12. 3rd Regiment in 1799.
1st to 4th Regiments of Horse, around 1768: 13. 1st.—14. 2nd.—15. 3rd.—16. 4th.—These four regiments were in turn converted to Dragoon Guards as the 4th, 5th, 6th and 7th Regiments (see figs. 18 to 21).—17. Rear view of a uniform in one of the Horse regiments. The white cartridge pouch, which had replaced the natural leather version, was itself replaced by a black one from 1793.
18 to 21. 4th, 5th, 6th and 7th Dragoon Guards (formerly the 1st, 2nd, 3rd and 4th Horse until their conversion to Dragoon Guards in 1789). Lawson shows the colours of the 4th as yellow and those of the 5th as green. Hats (not shown) were all bordered with braid of the same colour as the metal of the buttons.

The Light Cavalry

It is surprising to find that Britain was the last great power to incorporate hussar units into her army; they did not appear until 1805, and then only as 'Light Dragoons', three regiments of which were entitled to call themselves 'hussars' but appeared in the Army List as the 7th, 10th and 15th Light Dragoons with the word (hussars) in brackets. This long resistance to an innovation which had found universal favour elsewhere was no doubt the result of the insular attitude of the English authorities and their distaste for exotic uniforms.

There is a school of thought still widely represented today which maintains that the European hussar regiments were simply ordinary light cavalry in fancy dress who hoped to acquire some of the enviable reputation of the genuine Hungarian hussars, along with their uniform. The reader need only turn to the descriptions of the hussars in the French and Prussian chapters to see for himself how inaccurate this viewpoint is.

In 1728 a general by the name of Hawley had suggested the establishment of a light dragoon regiment equipped to fight in the traditional style of the mounted infantry, which methods as we have seen, had been steadily abandoned by the existing dragoon regiments. His proposal remained on the file, but one veteran of the European campaigns, the Duke of Kingston, became convinced of the absolute necessity of light horse during the Jacobite rising of 1745. At his own expense he raised the 'Duke of Kingston's Light Horse', 'to imitate the hussars in foreign service'.[1] Though the imitation was restricted to tactics and the adoption of the curved sabre, the new regiment was outstandingly successful – so much so that, once the revolt had been crushed and the regiment demobilized, the Duke of Cumberland asked and received permission from his father, George II, to re-establish it under his own command. But the Duke of Cumberland's Dragoons, numbered as the 15th Regiment on the dragoon list, survived only from September 1746 until February 1749; the reason for its suppression is unknown.

With the reopening of hostilities with France the need for light troops was once again apparent. At the end of 1755 each regiment of Dragoon Guards and the 1st, 2nd, 3rd, 4th, 5th, 7th, 10th and 11th regiments of dragoons were strengthened by the addition of one company of light dragoons. Interestingly, these troops were often known as 'hussars', though never officially. The excellent results they achieved encouraged the authorities to develop them into separate regiments. Five such regiments were added to the dragoon list in 1759, and some fifteen others were to follow in the next thirty-six years. To the beginning of the following list may be added the regiments formed with the help of the 7th, 8th, 9th, 10th, 11th, 12th, 13th and 14th Regiments of Dragoons.

GREAT BRITAIN – CAVALRY, MUSICIANS AND STANDARDS
1. Kettledrummer, 2nd Regiment of Horse, in 1751. The kettledrummers were normally mounted on grey horses. Unlike the trumpeters, who were long regarded as non-combatants, they had a duty to defend their instruments – highly prized spoils of war – to the last. – 2. Kettledrummer of the 1st Horse in 1751. – 3. Drummer of a dragoon regiment. The front of the mitre displayed either the standard motif shown here, or the regiment's own emblem, if it had one. The ribbon forming the back of the mitre was the same colour as the front panel and was embroidered, in the centre, with a drum and the regiment's number. The lining of the coat, like the front panel of the mitre, was in the regimental colour, the braid being a miniature version of that on the horse's saddlecloth. – 4. Trumpeter of a royal regiment around 1740.
Standards: 5. Guidon of the 1st Company of Horse Grenadiers. The 2nd Company had an identical one, but with a crimson background. – 6. Standard of the 1st Company of the Life Guards. – 7. Standard of the 2nd Company of the Life Guards. – 8. Standard of the Royal Horse Guards. – 9. Guidon of the 3rd Squadron of the 2nd Dragoon Guards. The 1st Squadron had a crimson one, the 2nd a blue.

---

[1] Captain R. Hinde, *Discipline of the Light Horse* (1778).

DATE RAISED	NAME	NOTES
1759	15th Light Dragoons	'Royal' from 1766: King's Regiment of Light Dragoons
1759	16th Light Dragoons	'Royal' from 1766: Queen's Regiment of Light Dragoons
1759	17th Light Dragoons	Disbanded in 1763
1759	18th Light Dragoons	17th from 1763; the 'Death or Glory Boys'
1759	19th Light Dragoons	18th from 1763
1760	20th Light Dragoons	Disbanded in 1763
1778	19th Light Dragoons (new)	Formed from the light companies of the 1st and 2nd Dragoon Guards and the 4th and 10th Dragoons; disbanded in 1783
1778	21st Light Dragoons (new)	Formed from the light companies of the 2nd, 3rd, 7th, 15th and 16th Dragoons; disbanded in 1783
1780	22nd Light Dragoons	Also known as 22nd Sussex; disbanded in 1783
1780	23rd Light Dragoons	19th from 1786; the first British cavalry regiment to serve in the Indies
1792	20th Light Dragoons (new)	Known as 'Jamaica Light Dragoons'
1794	21st Light Dragoons (new)	
1794	22nd Light Dragoons (new)	
1794	23rd Light Dragoons (new)	
1794	24th Light Dragoons (new)	
1794	25th Light Dragoons	Known as 'Gwyn's Hussars' or 'British Hussars'
1795	26th, 27th, 28th, 29th Light Dragoons	Formed from the supernumerary recruits of 3 regiments of Dragoon Guards, 2 of Dragoons and 7 of Light Dragoons
1795	30th, 31st, 32nd, 33rd Light Dragoons	Raised in Ireland in 1794, numbered in 1795 and disbanded in 1796

GREAT BRITAIN, DRAGOONS (I), 1742

1. 1st Regiment or Royal Dragoons. This was the only regiment to wear a lanyard on the left shoulder.—2. 2nd Regiment of the Royal North British Dragoons (later the Scots Greys). This was the only regiment to use grey horses, a custom which dated from 1694. It was also unique in having white leather equipment.—2a. Ceremonial mitre of the 2nd Regiment; in bad weather it was protected by an oiled cover. Under normal conditions the regiment wore the standard hat used by all the others. On the left is shown the authentic way the coat-facings were fastened, taken from a uniform which has miraculously survived until the present day; contemporary pictures never show this detail.—3. 3rd Regiment or King's Own Regiment of Dragoons.—4. 4th Regiment.—5. 5th Regiment or Royal Irish Dragoons.—6. 6th Regiment, the Inniskilling Dragoons.—7. 7th Regiment, or Queen's Own Regiment of Dragoons.—8. 8th Regiment. This regiment wore cartridge belts captured in Spain. Sword and bayonet were slung from a second cross-belt, unlike the other dragoon regiments and in the style of the heavy cavalry. Note the extra-large buttonholes and plain cuffs which also distinguish this decidedly non-conformist regiment. The regimental colour, shown here as orange, had been yellow in 1735 and would be again in about 1751.—9. 9th Regiment.—10. 10th Regiment.—11. 11th Regiment. This illustration shows the manner of wearing the sword and bayonet; the sling which supports them was an integral part of the sword-belt worn over the waistcoat. All regiments except the eccentric 8th used this system.—12. 12th Regiment.—13. 13th Regiment.—14. 14th Regiment.

Distinctive Colours of the Light Dragoons in 1793

REGIMENT	FACINGS	LACE	TURBAN	PLUME
7th	white	white	white	white (red base)
8th	red	white	red	red
9th	buff	white	red	red
10th	yellow	white	yellow	yellow
11th	buff	white	buff	buff
12th	buff	white	buff	buff
13th	buff	yellow	buff	buff
14th	orange	white	orange	orange
15th	red	white	red	red
16th	red	white	red	red
17th	white	white	white	white (red base)
18th	white	white	white	white (red base)
19th	very light yellow	white	pale yellow	pale yellow
20th	yellow	white	yellow	yellow
21st	yellow	white	yellow	yellow
22nd	red	white	yellow	yellow
23rd	pale yellow	white	pale yellow	pale yellow
24th	pale yellow	white	leopard-skin	pale yellow
25th	red	white	red	red
26th	green (1796: purple-blue)	white	purple-blue	purple-blue
27th	white	white	white	white (red base)
28th	yellow	white	yellow	yellow
29th	pale buff	white	pale buff	pale buff

The 8th, 19th, 25th, 27th and 28th Light Dragoons were permitted to wear a pale grey broadcloth uniform for tropical service. The regiments numbered 30 to 33 are virtually unknown, and even the briefest summary is not possible.
The white and yellow braid became silver and gold respectively in the case of officers.

GREAT BRITAIN, DRAGOONS (II), 1750–58
1. 1st Regiment (Royal Dragoons) in marching order. Note the tent-pole attached to the carbine. The coat-tails are tucked up behind over the rolled-up cloak. The same general outline is good for all regiments.—1a. Position of the epaulet and shoulder knot.—2. 2nd Regiment (Royal North British Dragoons). Still the only regiment with white leather. —2a. Ceremonial mitre. The interior was red, and ended in a white tassel; the back of the mitre was decorated as shown in fig. 17 below.—2b. Detail of the front panel.—3. 3rd Regiment (King's Own Regiment of Dragoons).—4. 4th Regiment. 5. 5th Regiment (Royal Irish Dragoons. This regiment was disbanded, rightly or wrongly, in 1797, for having recruited too many Irishmen who were suspect in the eyes of the British Government, and for being deeply involved in the Irish Rebellion. The number remained in abeyance for several years.—6. 6th Regiment (the Inniskilling Dragoons).—7. 7th Regiment (the Queen's Own Regiment of Dragoons).—8. 8th Regiment. The leather crossbelts had remained a regimental trademark, despite many attempts to get rid of them. Behind them, the buttonholes are still grouped in threes.—9. 9th Regiment.—10. 10th Regiment.

The buttonholes here are grouped three, four and five (reading from top to bottom).—11. 11th Regiment.—12. 12th Regiment.—13. 13th Regiment.—14. 14th Regiment. —15. Wreath of roses and thistles distinguishing the non-royal regiments.
16. Soldier of a light company (11th Regiment). The special cap carried a plume with a red base, the rest of it being in the regimental colour. The uniform was that of the regiment to which the light company was attached: 1st, 2nd or 3rd Dragoon Guards, 1st, 2nd, 3rd, 4th, 5th, 6th, 7th, 10th or 11th Dragoons. The smaller cross-belt supported the cartridge pouch, the larger the carbine; but the larger is also known worn on the other shoulder and carrying the cartridge pouch; or two large belts may be worn, one with the pouch and the other with the carbine. The light companies were disbanded in 1763.—17. Special headgear of the light company of the 2nd Regiment. The front part was identical to that of the mitre shown in fig. 2a, but less tall and thin.— 18. Uniform of regimental farrier (here, the 11th Regiment). The axe, carried in a little case slung from a crossbelt, was used to finish off hopelessly injured horses. Note the leather apron, rolled up round the waist for riding. It seems likely

that these specialist troops always wore a blue uniform, with the distinctive buttonholes, etc., of the regiment in which they served. The holsters were replaced by two cylindrical black leather cases containing horseshoes, nails, etc. The farriers horses were not necessarily the same colour as those of the regiment in general.—19. 1st Regiment (Royal Dragoons) in 1768. The ordinance of 1768 retained the same regimental colours but introduced a turn-down collar, smaller cuffs, and a new style of hat with a less exaggerated peak in front and sometimes with tassels at the sides, according to the very rare contemporary sources. Plumes seem to have made an appearance, contrary to regulations (they were not officially allowed until 1799). Another significant result of this ordinance was the adoption of white cloth for waistcoats, breeches and coat-tail facings, except in the 3rd, 9th, 11th and 13th Regiments, where all these were buff-coloured. The general outline of fig. 19 applies to all regiments except the 2nd Dragoons, which had two epaulets.—20. 1st Regimental (Royal Dragoons) in 1799. Once again, we can see the extraordinary transformation which the British uniform underwent during the very last years of the century. The imaginative (and hence subjective) reader will be unable to resist the thought that this was, as it were, the prologue to the great drama: the actors have changed their costumes, and are waiting for Destiny to call them.

GREAT BRITAIN, LIGHT DRAGOONS (pp. 58–9)
1. Cumberland Dragoon in 1747. All details of uniform and equipment are taken from a particularly meticulous picture by David Morier (see Vol. I, p. 108). Red and green were the colours of the Duke of Cumberland's livery. The plume, ahead of its time, disappeared with the regiment itself in 1748; it may have had a red base. The royal crest (the crowned lion of England) on the saddlecloth was topped by the royal crown, as on the holster-covers. The saddlecloth is shown a little over scale here, for reasons of clarity. Its overall height should be reduced by the width of the braid round the edge.—2. 16th Light Dragoons. On being made a royal regiment in 1766, the 16th took blue as its new regimental colour, and used it for its coat facings. Apart from the loss of the plume and the adoption of fringed epaulets, the general outline remained unchanged (see fig. 3).—3. 16th (Queen's) Light Dragoons in 1766. The saddlecloth carries the monogram C for the Queen, Sofia-Charlotte von Mecklenburg-Strelitz. The one-piece shabraque is bordered with the braid reserved for the royal regiments.—4. 15th Light Dragoons. Until 1767 or thereabouts the front panel of the cap bore the surprising legend: *Five battalions of French infantry defeated and taken by this regiment, Emsdorf, July the sixteenth 1760.* Below the royal crest, in the centre, is the motto *Merebimur*, with the captured flags to right and left. This excessive display of complacency was no doubt at the root of the change in 1767 to the new headgear shown in fig. 14. The old regimental green gave way to blue in 1766, the regiment being elevated to 'royal' status at the same time as the 16th (see fig. 3).—5. 1st Light Dragoons in 1763 (formerly the 18th, raised in 1759, see text). The death's head and the black line running through the braid were adopted in memory of the death of General Wolfe before Quebec in 1758; the colonel of the regiment, Hale, had served with Wolfe in Canada. The scroll under the skull shows the words *or Glory*.—6. Drummer of the 15th Light Dragoons, 1760. The saddlecloth was of the type shown in fig. 2 but with a green background (this being the regimental colour until 1766), the regimental number being in white on a red background, surrounded by the traditional wreath of roses and thistles. The horse's trappings were bordered with white braid with a red central line.—7. 16th Light Dragoons in the new uniform jacket of 1784—this was worn below a second, sleeveless jacket, called a shell (see next figure).—8. 16th Light Dragoons around 1763, with shell. The arm-holes which had had 'wings' added in 1790, produced the illusion that the sleeves were part of the shell, whereas (as shown in the previous figure) they in fact formed part of the under-jacket. The red 'turban' was replaced by a black one towards the end of the century, a practice adopted in several other regiments. The 10th and 24th adopted a leopard-skin version at about the same period. The horse's trappings were in one piece and could do duty as a blanket. The front part bears the royal monogram GR, with a crown and the name of the regiment; the rear displays the Queen's monogram (see fig. 3) in the centre of the Garter.—9. Officer of the 15th Light Dragoons in 1796. At this time the horse's trappings were reserved for ceremonial occasions.—10. Sergeant of the 10th Light Dragoons in 1793. He wears two horizontal chevrons on his sleeve, though these marks of rank had not yet been officially recognized. The central line of the saddlecloth border is sometimes shown as blue edged with yellow instead of yellow, edged with blue, as here.—11. Officer of the 7th Light Dragoons in 1786. His coat, though very impressive to look at, is in fact out of date; based on the model prescribed in 1768, it was permitted as off-duty wear.—12. Officer of the 7th Light Dragoons in 1796.—13. Officer of the 10th Light Dragoons in 1796. The Dutch campaign of 1794 led to the introduction in Britain of the mirliton worn by the émigré and foreign troops encountered abroad; at the same time the uniform of 1784, whose so-called 'hussar pattern' had been rather too approximate, was now able to receive more authentically hussar-like embellishments. 14. Helmet of the 15th Light Dragoons, 1767. This model replaced the one shown in fig. 4. Only the captured trophies and the word 'Emsdorf' were retained, one either side of the central motif. The elegant 'classical' helmet, still worn in 1788 and pronounced out-of-date—reasonably enough, it must be admitted—by the authorities was now abandoned. The only known example is in the magnificent collection of R. and J. Brunon in the château de l'Emperi at Salon-de-Provence.—15. Helmet of the so-called tarleton pattern, dating from the end of the eighteenth century. It belonged to an officer of the 9th Regiment. The emblem is particularly fine.—16. Uniform of an officer of the 11th Light Dragoons in 1799. Note the increased quantity of braid, already noticeable in No. 12, and the 'hussar' effect thus produced.—17. Helmet of the 17th Light Dragoons (see also fig. 5), 2nd (1778) model.—18. Helmet of the 20th ('Jamaica') Light Dragoons. Not to be confused with the original 20th Regiment, raised in 1760 and disbanded in 1763 (see text). The cayman emblem was retained until 1802.—19. Helmet of the 21st Light Dragoons in 1760. Despite the proud device *Hic et*

The Horses

Apart from the early dragoon regiments, the British cavalry were always mounted on powerfully built horses. In 1729 the ideal height at the withers was laid down as 15.1 hands for horse regiments and 15 hands for dragoons, the insignificant difference being accounted for by the fact that the dragoons had already abandoned their traditional function as mounted infantry to become genuine heavy cavalry. The light dragoons, on the other hand, were from the start mounted on horses of 14.3 hands.

The impression of power given by these horses was strengthened by their black coats, in the cases of the Life Guards, Royal Horse Guards, King's Dragoon Guards, Royal Dragoons and King's Dragoons: black horses always appear stronger than they are.

The Queen's Dragoon Guards used bay horses, and the Royal North British Dragoons their famous greys; all the other heavy cavalry regiments were mounted on horses of some shade of brown or chestnut.

The light dragoons always preferred black horses, or horses that would pass for black.

Buglers were always mounted on greys.

Tails were cropped extremely short until 1764, in which year it was officially decided that tails should be of their natural length; for obvious reasons this ruling was some time in taking full effect. From 1799 tails were once again cropped.

ubique ('Here and everywhere') which may be seen below the royal monogram, this regiment led an exceptionally peaceful existence on British soil between its foundation in 1760 and its disbandment in 1763. The strangely asymmetric front panel is based on a Hanoverian model. The letters RF stand for its nickname (Royal Foresters).—20. Tropical helmet, 8th Light Dragoons, 1796—some regiments saw service in the colonies. A forerunner of the heavy cavalry helmet, not adopted until 1812 (see *Arms and Uniforms of the Napoleonic Wars*). Helmets had a 'turban' tied at the back by a loose bow; undoing the knot provided the neck with protection against rain, and must have looked sensational!

The Artillery

British artillery of the early eighteenth century consisted of a wide variety of weapons officially distinguished by the weight of the shot, in the case of cannon, or the calibre, in the case of mortars. The most severe problem was getting this cumbersome collection to the battlefield. A standard siege train of some 160 guns required a convoy of 3,000 wagons and 15,000 horses, proceeding at walking pace and extending for some fifteen miles![1] The civilian personnel manning the train abandoned their guns at the first hint of danger, and it was necessary to raise a special regiment for the purpose in 1685: the Royal Fusiliers, though the regiment soon lost its exclusive artillery status and was known from 1690 onwards as 'the 7th Regiment of Foot, Royal Fusiliers'.[2]

The artillery train included every conceivable type of specialist, from engineer to blacksmith, but it existed only as long as each campaign lasted. As soon as peace returned it was immediately dissolved. Not until 1716 was a specialist corps formed on a permanent basis: two companies strong to begin with, it was increased to eight by 1744; further expansion saw the 'Royal Regiment of Artillery' organized in two battalions by 1757, and in four by 1771.

---

[1] Marlborough was the first general to 'export' a complete artillery train to Europe, but he had to manage with 34 cannon once and for all.

[2] For uniform, see Vol. I, p. 93, fig. 7.

Throughout this period the guns themselves were constantly being improved. On the theoretical side Benjamin Robins's *Principles of Gunnery* (1742) laid the foundations of the scientific study of artillery practice, and Thomas Binning's *A Light of the Art of Gunnery* (1744) was of great practical value. But Henry Shrapnel's revolutionary invention of the 'ball shell' in 1784 was not exploited until the beginning of next century, in 1803.

The artillery uniform, usually red until 1716, became blue with red facings.

The Royal Horse Artillery, established in 1793, was soon known everywhere as the 'Flying Artillery' or 'Galloper Guns', which did justice to its astonishing mobility. Their original uniform – the same as the rest of the artillery, but with cavalry boots – was soon replaced by the hussar style of the light dragoons.

Finally, a word about the Honourable Artillery Company. Regarded as the oldest regiment in the world, its origins go back to the Guild of St George, founded during Henry VIII's reign in 1537 and developed as an 'artillery company' responsible for trying out all the long-range weapons of the period. This admirable nursery for artillery officers acquired the title 'Honourable Artillery Company' in 1668. Its first Captain General was the Duke of York in 1682, since when most British sovereigns, up to the present day, have borne the title. The H.A.C. provides the Lord Mayor of London with the escort of pikemen that accompanies him on ceremonial occasions. It collected thirty-eight citations for gallantry during the First World War, and fought in North Africa, Sicily, Italy and the west during the Second.

GREAT BRITAIN, ARTILLERY AND ENGINEERS
1. Bugler and drummers of the Royal Artillery at the time of the Seven Years' war. Both wore coats very similar in decoration to their counterparts in the cavalry (see plate facing p. 52); and they too displayed the crowned royal monogram on back and front. Note the unexpected position of the driver's cockade.—2. Flag of the Corps of Artificers in 1760.—3. Honourable Artillery Company sergeant in 1725 (see text)—4. Artillery officer in 1742.—5. Artilleryman in 1742.—6. Soldier of the artillery train in 1756.—7. Artillery officer in 1760.—8. Artilleryman in 1797. Note the pricker on his bandolier, used to unblock the aperture of the cannon.—8a. Cartridge pouch bearing the insignia of the artillery (since 1770 or earlier)—a crown on a red cloth background, above a banderol with the designation R. ARTIL. followed by the battalion number in arabic figures.—9. Engineer officer in 1778.—10. Artillery officer in 1788. The gold braid disappeared from the uniform soon after this date. Change the boots for lighter, calf-length ones, and add a white plume to the hat, and you have the general outline for the last decade of the century.—11. Engineer officer in 1782. The cut of the uniform is virtually identical to the red-coated version of 1778 (fig. 9).—12. Engineer officer in 1792. The artillery too adopted the top-hat in one form or other, of varying shapes.—13. Engineer officer in 1797.

PRUSSIA, SAXONY, BAVARIA AND OTHER GERMAN STATES

In our first volume we examined the birth of the Prussian army, the monarchs who worked to establish it, the organization of its infantry, the system of discipline that gave it its massive strength. We can now add a series of chapters introducing the renowned Prussian horsemen of the eighteenth century: hussars, dragoons, cuirassiers, uhlans; and a brief survey of the Prussian artillery. The rest of this chapter is devoted to other troops of the Holy Roman Empire, some – like those of Saxony and Bavaria – earning a section to themselves, some simply shown in two pages of drawings.

The Hussars

The first hussars appeared in Prussian service in 1721. By 1735 they were known generally as 'Prussian hussars' to distinguish them from a second body established in 1731 and known as the 'King's' or 'Berlin' hussars. Under Frederick II the two corps were reorganized as regiments, the first taking the name of Bronikowski and the second that of Zieten. Rather than designating the regiments shown in our illustrations by the names of their successive commanding officers, which would have involved captions of almost impenetrable complexity, we have adopted the numbering system introduced in 1806, based on order of seniority.

The term *Chef*, roughly equivalent to Colonel in Chief, does not imply that the holder of that rank, usually a general, commanded the regiment in the field; he normally delegated this role to a *Kommandeur* or *Kommandant*, an officer with the rank of major or lieutenant colonel.[1]

PRUSSIA, HUSSARS (I)
In this and the two following pages, each group of uniforms shows, from left to right, the dolmans of a trooper, an N.C.O., a bugler and an officer.
1. 1st Regiment: a. Dolman from 1721 to 1732.—b. Dolman from 1732 to 1742.—c. Trooper's shabraque.—d. Officer's shabraque.—e. Officer's parade sabretache.—f. Variant.—g. Officer's regular-service sabretache, and pelisse.—h. Trumpeter's dolman braid and surrounding trim.—i. Officer's kolpak.—j. Trooper's frogging (18 rows for each man).—k. Hussar of the 1st Regiment in 1762, the year in which the plume was adopted by all regiments. The heart-shaped cut-outs in the thigh-length scharawades disappeared early in the Seven Years' War (1756–63). Up until 1740, these curious items of uniform were dark blue in both the hussar units then existing, the Berlin Hussars and the East Prussians, both raised by King Frederick William I, father of Frederick the Great.—l. Hussar of the 1st Regiment in 1798. The shako was not introduced until 1806.
2. 2nd Regiment: a. Trumpeter's dolman and pelisse.—b. Frogging (18 rows) and edging trim.—c. Trumpeter's mirliton.—d. Officer's sabretache.—e. N.C.O.'s mirliton.—f. N.C.O.'s dolman and pelisse (detail).—g. Officer's ceremonial or dress sabretache.—h. Officer's shabraque.—i, j, k. Hussar (his pelisse was trimmed with white fur), N.C.O. and ensign. Note the braid (white for the trooper, silver for the N.C.O. and gold for the officer) which surrounded the frogging on both dolman and pelisse.—In the centre of the page is our impression of the famous Hans Joachim von Zieten, known as the 'father of the Prussian hussars', his facial details being based on a portrait by Therbusch painted in 1769. The uniforms shown are in the colours worn between 1732 and 1807; from 1730 to 1731 the dolman was white with dark blue collar and cuffs, then light blue with red collar and cuffs.
3. 3rd Regiment: the figure on the left is a trumpeter.—a. Trooper's shabraque.—b. Officer's shabraque.—c. Officer's shabraque (variation).—d. Trooper's sabretache.—e. Regular service and parade sabretache.—Frogging from dolman (18 rows).

---

[1] As in the case of von Seydlitz in the cuirassiers (see p. 76).

The Father of the Prussian Hussars

For an example of the rank of *Chef* we need look no further than the most famous of all, Joachim von Zieten, who took over the 2nd Regiment in 1741, during the First Silesian War,[1] and retained it until his death in 1786. A born soldier, Zieten began his military career as a foot soldier, at the age of fourteen. He soon left the army, but re-entered it as a twenty-seven-year-old lieutenant of dragoons. He then quarrelled with a superior officer and was sentenced to a year's imprisonment and loss of rank. Rehabilitated in 1730 he joined the 2nd Berlin Hussars, then known as Beneckendorf's.

By 1735 he had reached the rank of major, but his contribution in the First Silesian War was so outstanding that the year 1741 saw him a colonel, and *Chef* of his regiment. In 1744, during the Second Silesian War, he was promoted Major-General, and fought with distinction at Hohenfriedeberg and Katholisch-Hennersdorf, being seriously wounded at the latter. This was the start of a disastrous period in the life of this brilliant officer. He lost first his wife, then his son, while envious colleagues took advantage of his absence to discredit him in the eyes of the King he had served so well.

But the Third Silesian War, or Seven Years' War as we know it, saw Zieten's return to favour. Offered the chance to deploy his talents to the full, he carried off the honours at Reichenberg, Prague, Leuthen[2] and Torgau.[3] Promoted again, this time to General of Cavalry, this most popular of all the great cavalry commanders of the time had to be dissuaded by the King himself from taking an active part in the War of the Bavarian Succession; he was then eighty years of age. He died in 1786, only seven months before Frederick II himself.

---

[1] German historians recognize three Silesian wars: the First, from 1740–2; the Second, 1744–5; and the Third, which we know as the Seven Years' War, 1756–63.

[2] See Vol. I, p. 142.

[3] Where, two centuries later, the American Third Army under Patton joined up with Koniev's Soviet troops.

PRUSSIA, HUSSARS (II)

1. 4th Regiment: Their white pelisses sometimes led to these 'white hussars' being known as *die Schafe* ('the sheep'). —a. From 1798–1807. The following series refer to the period 1752–87.—b. Trooper's frogging (15 rows).—c. Trumpeter's frogging and braid from 1787.—d. Belt, up till 1752.—e. Belt, 1787–1806.—Until 1752 the scharawades (also called schalawarys or scharivaris) were white.

2. 5th Regiment: a. Dolman from 1771 onwards.—b. Trooper's shabraque.—c. Officer's shabraque.—d. Trumpeter's rosette, from 1771.—e. Trumpeter's plume from 1771.—f. Frogging (12 rows). The buglers' frogging was changed to red and white from 1771.—Left, the mirliton. The insignia was worked entirely in white thread, with the eye sockets, nasal cavity and lines between the teeth in brown thread. An ancient tradition maintains that these skulls were taken from the funeral hangings of Frederick William I. This emblem earned the regiment the nickname of *Totenköpfe* ('death's heads'); they were also known as the Black Regiment.

3. 6th Regiment: Known as the *Fleischhacker* ('butchers') or sometimes *Kapuziner* ('capuchins'—an allusion to the similarity of the colour of their uniform to monastic home spun). —a. Regulation shabraque and pack: at the bottom, the oatmeal bag; centre, the saddle bag; top, rolled-up cloak. Despite its name, the cavalry portmanteau in the eighteenth century was more likely to contain cutlery and assorted linen than a cloak.—b. Officer's shabraque.—c. Officer's pelisse.—All ranks wore 15 rows of frogging.

4. 7th Regiment: Known from the yellow colour of their dolmans as the *Kanarienvögel* ('canaries').—a. Dolman from 1771 to 1807.—b. Trumpeter's frogging (all ranks had 12 rows).—c. Trooper's shabraque.—d. Officer's shabraque.—e. Officer's pelisse.

5. Hussar of the 4th Regiment in pelisse.—6. Trumpeter of the 5th Regiment. Note the absence of a carbine sling—trumpeters were not armed with this weapon. Some sources show the *scharawades* as black, the same as the trooper's.—7. Officer of the 6th Regiment. The fur trimming of the pelisse was changed to white in 1771.—8. Hussar of the 6th Regiment.—9. Officer of the 7th Regiment in pelisse.—10. Hussar of the 7th Regiment.

Hungarian Beginnings

No man with Frederick William I's eye for a good soldier could have been unimpressed by the sight of the first Hungarian hussars, driven into exile after Rakoczy's rebellion.[1] It seems, however, that the hussar regiment Frederick William established contained a fairly high percentage of Germans, unlike those in the French service, which had been able to draw on successive waves of Hungarian refugees arriving in France ever since the early seventeenth century.

Kolpak and Mirliton

The original hussar headgear was the *mirliton*, a Hungarian cap shaped like a truncated cone. Curiously enough this 'typically Hungarian' headgear is in fact regarded by Hungarian historians as a French innovation, introduced under the Angevins in the fourteenth century. The true Hungarian hat was a fairly tall, cylindrical cap of animal fur (wolf, sheep or fox) called the *hajduk*, with a loose lining (the *flamme*) usually of the same colour as the dolman.

In Prussia, however, the first hussars wore the *mirliton*. In 1732 it was replaced by a red-lined cap trimmed with fur. From modest beginnings this fur grew in a few years to a height of a foot or so, and by 1740 it could fairly be described as a true *kolpak*.

Of the two, the *mirliton* was certainly the more popular, at least during the summer months, from 1741 onwards. In that year Frederick II ordered Colonel von Massow to have the *mirlitons* for the 5th Regiment designed in imitation of one taken from an Austrian pandour. The Berlin hatter charged with this task decided, out of curiosity, to unwind the long *flamme*, and found painted on the inside a death's head, the existence of which had not previously been noticed. Told of this depressing discovery, Frederick decided to adopt it, but 'Prussia will display openly what Austria conceals'.

The origins of the death's heads, embroidered in white thread, are obscure. One theory is that they were cut from the trappings used at the funeral of Frederick the Great in 1740. Another,

PRUSSIA, HUSSARS (III)
1. 8th Regiment (12 rows of frogging): This regiment was disbanded at the surrender of Maxen in 1759, and officially ceased to exist at the end of the Seven Years' War, when its number was reallocated to von Belling's regiment (fig. 2).
2. 8th Regiment (18 rows): Raised as von Belling's Battalion of Hussars in 1758, and reorganized as von Belling's Regiment in 1761. The regiment was nicknamed '*ganze Tod*' ('complete death') from its emblem, a complete skeleton, by way of contrast to the 5th *Totenköpfe* Regiment. The black uniforms shown date from the period 1758–64. In 1764 the regiment took over the number and the uniform of the 8th (see 2a, 2b and 1), and so became the New 8th, destined to find fame under the command of Blücher.—a. Uniform from 1764 to 1789.—b. Uniform from 1789 to 1807.—c. Trooper's shabraque.—d. Officer's shabraque.—e. Mirliton (details). —It should be mentioned that according to the great Menzel the trooper's sash was green with yellow loops and cord (tie-cord), while other modern German sources show the whole thing as black.
3. 10th Regiment: (15 rows):—3a. N.C.O. It will be remembered that the plume was introduced for all regiments in 1762.—3b. Trumpeter, with (left) a detail from his special braid.
4. Bosniak lancer, N.C.O. Surprisingly enough this regiment is not out of place here—it was numbered as the 9th Hussars. More decorative than useful, this type of soldier was one of many instances of the Turcophilia rampant at the time. Other ranks had no buttonholes on the caftan, and their cuffs were trimmed with the narrow braid used by the hussars proper.
5. 11th Battalion of Hussars (18 rows): Raised in 1792 under the reign of Frederick William II.—5a. Hussar. From 1787 onwards the Prussian hussars wore white cloth breeches, and over them the overbreeches shown in fig. 5b.—5b. Hussar in pelisse and overbreeches; the monogram of Frederick William II is visible on the sabretache.

---

[1] See, 'The Hussars', p. 112.

equally plausible, maintains that they were loot from a sacked Silesian monastery whose inmates specialized in making funeral hangings.

As late as 1914 the significance of these emblems was persistently misunderstood abroad: they were intended to symbolize steadfastness unto death, refusal to surrender, instead of which the 'Death's Head Hussars' merely acquired a sinister reputation for giving no quarter.

There was no such ambiguity about the device favoured by the 8th Regiment from 1758: going one better, they adopted a complete skeleton armed with a scythe and the motto *Vincere aut mori* – 'Conquer or die'. Both the 5th and the new 8th Regiments seem to have been so proud of their gruesome emblems that they rarely wore the *kolpak*.

But the brown *kolpak* was certainly worn, at least in winter, by the regiments numbered 1, 2, 3, 4 and 10, who kept it until 1796. Dark brown fur was favoured in the 2nd, 10th and old 8th Regiments. Only the trumpeters remained faithful to the *mirliton*, which in fact came back into more general favour in the first years of the nineteenth century, until the increasing change to the shako between 1804 and 1806.

In the 2nd Regiment (Zieten's) the officers' *kolpaks* were embellished, from 1743 onwards, with a sort of gilt sceptre, from which flew a small black eagle's wing. Subalterns were restricted to a white tuft of feathers sprouting from a short black plume. These plumes were worn in all the other regiments after 1762; in the case of a *Chef* of general's rank the plume would be of ostrich feathers with a black base.

The officers' *mirlitons* were at first bordered with silver or gold; by 1790 this metal had spread to cover the whole inside surface of the *flamme*, and later the outside surface too, both then being trimmed with a fine fringe of black silk.

The Hat

The wide tricorn known as the 'German hat' was part of an officer's undress uniform, and was also worn by off-duty officers. From 1797 this latter was its only function, and then only for officers below the rank of general.

The Scharawades

Over his characteristic tight leather breeches the hussar wore curious, stocking-like garments called *schalawarys*, *scharivaris* or *scharawades*, which had the design of a heart cut out on each thigh. These disappeared entirely from use in 1787 with the adoption of overbreeches.

PRUSSIA, DRAGOONS (I)
Figures 1 to 6 show the uniform, shabraque and pompom of the first six dragoon regiments from 1746. On the right of each figure is shown the officer's button-hole which appeared on the cuffs, on and under the coat facings, on the pockets, and at the back on the buttons at the waist (the 5th Regiment wore them only at the back and under the coat facings, the 6th under the facings and on the pockets). See the following page, 5th and 6th Regiments.
7. Horse Grenadier in 1730, in the reign of Frederick William I. This regiment was disbanded by his son, Frederick II, at the beginning of the First Silesian War in 1740, for having almost managed to lose the King to the Austrians when serving as his personal escort. — 8. Dragoon of von Schorlemmer's Regiment (the 6th) in 1745. This unit was known as the *Porzellanregiment* ('porcelain regiment') when the Soldier King exchanged it for a collection of jade and Chinese porcelain which had belonged to his father, the luxury-loving Frederick I. With impeccable logic, the collection used to make this astounding transaction were known at the court of Dresden as the *Dragonervasen*. — 9. Dragoon of the Bayreuth Regiment (the 5th) at the time of the First Silesian War (1740–2).
10. Steel skull-caps worn into battle beneath the tricorn hat, a practice followed by all mounted troops of the time who wore the hat. There were many different varieties, more or less solid in design. — 11. Trooper's cross-belt, sabre and cartridge-pouch. — 12. Drummer's sword and belt, with dragoon's musket and tent-pole (see fig. 3).

The Boots

Hussars' boots were knee-length in front, rather shorter behind; they had no leather heel, but instead a horseshoe-shaped piece of iron called a *potkowke*. The leather heel came in around 1785, and did not become standard until the very end of the century. The bright yellow officers' boots survived only as a special privilege in Zieten's Regiment (the 2nd), for parade dress. The upper rim of the boot was always trimmed with gold or silver, according to the metal of the uniform buttons.

The Belt

The hussar's belt was another characteristic eccentricity. It consisted of a very long rope of twisted woollen yarn held together by sliding loops of very tightly plaited thread. It was passed four or five times around the waist and fastened at the back by a button and loop which were extended by a double tasselled cord, the *fouet*, so called because it resembled a whip. This was looped back across the flank and passed under the belt itself in such a way that its ends hung loose on the right thigh. Both officers and men at first wore the sliding loops arranged chess-board-fashion, but from 1797 onwards they were placed one above the other.

Weapons

The sword with a plain iron guard replaced one with a copper guard which had been in use before 1740. Around 1787 the eighteen-inch flintlock pistol gave way to a more modern type with conical touch-hole and iron ramrod. The hussar's arsenal was completed by a carbine, of variable length and style.

The ten 'carabineers' attached to each squadron were equipped with a special rifled gun, with iron ramrod, longer than the normal hussar carbine but also more powerful and accurate.

---

[1] See France, 'The Light Troops', p. 40.
[2] Petrovaradin, on the Danube in Yugoslavia. Here on 5 August 1716, during the Austro-Turkish War of 1715–18, a victory was won by Prince Eugène of Savoy-Carginan. Born in Paris in 1663, exiled to Austria, Prince Eugène was the most brilliant soldier of his generation.

Frederick William II, who ruled Prussia from 1786 until 1797, increased the carabineer section from ten men to twelve and armed them with a new rifled carbine only thirty inches long.

The Panther Skin

This was the exceptionally ostentatious privilege of the officers of the 2nd Regiment. They wore it instead of the usual pelisse, or fur-lined cloak, with dress uniform and at the grand spring review. The original thirty-two 'tiger skins' were presented to Joachim von Zieten and his officers after the victory of Rossbach;[1] they were probably captured from the Turks after the Battle of Peterwardein[2] in 1716, as these highly prized skins were the uniform of the Grand Viziers' Guards.

PRUSSIA, DRAGOONS (II)
Figs. 1–6 show the 7th, 8th, 9th, 10th, 11th and 12th Regiments with the officer's distinctive buttonholes (for positioning of these see fig. A).
7. Dragoon, 8th Regiment.—8. Ensign, 7th Regiment.—9. Officer of the 5th Regiment with gala shabraque. The undress shabraque and holster-covers were lacking the royal emblem.—10. Dragoon, 5th Regt, rear view.—11. Drummer of the 5th Regiment in 1762. The plume was introduced towards the end of the Seven Years' War.
12. From left to right, braid worn by the drummers of the first twelve regiments (except the 5th for which see fig. 11).

The Dragoons

At the beginning of his reign, in 1714, the Soldier King Frederick William I had four regiments of dragoons dressed in white uniforms with cherry-red trimmings. In 1727 they were issued with buckskin *kamisoles*, which were in turn replaced by straw-coloured cloth coats in 1733. The iron breastplate was worn until 1718.

The dragoons all wore the tricorn hat, except for Derfflinger's Regiment, which favoured a low mitre topped by a copper flash and bearing on the front the star of the Order of the Black Eagle. This regiment took the title of Horse Grenadiers.

A fifth regiment, later to win fame as the Bayreuth Regiment, appeared in 1717, to be followed almost at once by another, the extraordinary 'Porcelain Regiment', exchanged by the Prince Elector of Saxony for a collection of fine Chinese porcelain. Its nickname was misleading, and it distinguished itself during the Rhine campaign from 1734 onwards, and again later in the service of Frederick the Great.

The 7th, 8th, 9th and 10th Regiments were raised in the same year; the 11th followed in 1741 and the 12th in 1742, in the reign of Frederick II.

In 1745, at the end of the Second Silesian War, the dragoons acquired their light blue coats. Towards the end of the century lapels were fastened from collar to waist, and tunics accordingly disappeared. The edges of all the facings were the same colour as the rest of the coat, but trimmed with the distinctive regimental colours.

In 1797 the dragoons wore leather straps crossed on the chest.

The Cuirassiers

The number of regiments of cuirassiers had always been constant at twelve, and was not increased until shortly after the accession of Frederick II, whose new Guards were numbered as the 13th Regiment in 1756. On parade or manoeuvres the Guard took precedence, followed by, in order, the *Gendarmen* (10th Regiment), the *Leibkürassiere* (3rd), the *Leibkarabiniere* (11th) and then the other regiments in numerical order – 1, 2, 4, 5, 6, 7, 8, 9, 12.

In the early eighteenth century the cuirassiers' uniform was of soft leather, replaced by straw-coloured cloth around 1735, except for the 2nd Regiment, which wore bright yellow until 1806. The tunics, known as *chemisettes*, adopted the distinctive regimental colours, with decorative braid, except for the gendarmen (10th Regiment) who took to dark blue – previously the 10th had worn blue tunics, the others white.

PRUSSIA, CUIRASSIERS, 1756–63 (I)
1. 1st Regiment. This first drawing, less schematic than the others, shows the padding and lining of the cuirasse and the belt worn over the *koller* (jerkin). It also shows the relative proportions of the different items, and of the pompoms worn at the side of the hat. Below, special cuffs—N.C.O. on the left, officer on the right.—2. 2nd Regiment.—3. 3rd Regiment.—4. 4th Regiment.—5. 5th Regiment—the blue of the regimental colours, a very pale shade, was known as 'deathbed blue'.—7. The 'gens d'armes' regiment (10th) in 1735. From this date, the yellow leather of the *koller* began gradually to be replaced by straw-coloured cloth, which in turn steadily paled until by 1806 it had become white.—8. Ensign of the 8th Regiment. The shaft of the standard was painted in the regimental colour and fitted with metal rings.—9. Trumpeter of the 5th Regiment.—10. 2nd Regiment in 1756. The braid on the hat was dispensed with during the Seven Years' War. Note the carbine, with tent stake in place.—11. Officer of the 3rd Regiment. An officer's cuirasse was edged with a frill instead of padding; the belt round his waist was silver striped with black.—12. Cuirassier of the 4th Regiment. The carbine-sling, not shown here, was edged with the regimental colour, black in the case of the 4th.—13. Left to right, trumpeter's braid for the 7th, 8th, 9th, 10th, 11th and 12th Regiments.

The cuirass itself was of blackened iron, except in the case of the Guards, who wore theirs polished. The backplate had already disappeared in the days of the Soldier King, and under Frederick William II the breastplate in turn was abandoned. The most remarkable feature of the cuirassiers' uniform was the sabretache, borrowed from the hussars and worn very high on the thigh.

There were plenty of colourful characters in these regiments. Friederich-Wilhelm von Seydlitz-Kurbach (1721–73) was one of Frederick II's most famous officers, and the most brilliant cavalry general of his day. A major at eighteen, he effortlessly earned promotion after promotion, and in 1753[1] obtained command of the 8th Regiment of Cuirassiers, which he turned into a model unit. This magnificent regiment, which later bore the names of von Pannewitz (1776–87) and von Heising, was virtually annihilated in combat against Napoleon's armies in 1806.

Quite apart from his heroic conduct on most of the major battlefields from Hohenfriedeberg in 1745 to Freiberg in 1762, by way of Kolin and Rossbach,[2] Seydlitz reorganized the ponderous Prussian cavalry, and made of it a fighting machine to match the celebrated infantry. When he died, with the rank of Inspector General of the Silesian Cavalry, Frederick had a marble statue erected in his memory in a Berlin square.

Another cuirassier, this time from the *Leibgarde*, better known for his unofficial exploits, was the extraordinary Baron von Trenck. He first found favour with the King of Prussia, then was disgraced and imprisoned. He escaped and survived an incredible series of adventures across the length of Europe and finally capped a distinguished career by dying beside André Chénier on the Paris scaffold.[3]

---

[1] Seydlitz did not become *Chef* of his regiment until 1757, until which date the regiment was known by the name of von Rockow.
[2] Hohenfriedeberg is now Dabriomerz, west of Wroclaw in Poland, where Prussia beat the Saxons and Austrians. Freiberg, see Vol. I, p. 152. Kolin, see Vol. I, p. 142. Rossbach, see above, p. 40.
[3] The major European television services have combined to produce an excellent serial about this adventurer, which is something of a rarity in being very good on the subject of uniforms.

PRUSSIA, CUIRASSIERS 1756–63 (II)
1. 7th Regiment.—1a. Royal monogram.—2. 8th Regiment.—3. 9th Regiment.—3a. Detail.—4. 10th Regiment.—5. 11th Regiment.—6. 12th Regiment.—7. Officer of the 1st Regiment.—8. N.C.O. of the 6th Regiment. Note the black and white pompoms of his rank—9. Trumpeter of the 2nd Regiment, nicknamed the *Gelbe Reiter* ('yellow riders'). Trumpeters wore dangling false sleeves, one of which can be glimpsed here between his elbow and his chest. Modern authorities show seven rows of horizontal braid on each sleeve (see fig. 9 on the previous page).—10. Cuirassier of the Saxe-Weimar regiment in 1786. Under the successors of Frederick II hats grew taller and taller, coat-tails shorter and shorter. But the distinctive regimental colours were left unchanged until 1810.—11. Ensign of the Guard (13th Regiment) in 1770. Unlike the standards of the other cuirassier regiments, which were fastened to a shaft at one side, those of the 13th were in the form of an ancient Roman *vexillum*.—12. Officer of the 13th Regiment in gala surcoat, about 1797 (the back of the coat repeated the design of the front).—13. Trumpeter's braid, 1st, 2nd, 3rd, 4th, 5th and 6th Regiments.—14. The cuirass, showing the method of fastening.

The Freikorps Cavalry

We have already studied the infantry units of the *Freikorps*,[1] the irregular troops raised in Prussia during the Seven Years' War. Their cavalry was for the most part hussars – it was entirely natural for these volunteers to favour the most original and colourful uniforms available, but in any case their role as skirmishers was absolutely in the hussar tradition.

One of the finest and most impressive of these volunteer units was the *Freikorps* of Colonel Friedrich Wilhelm von Kleist, who had been *Chef* of his own regiment, the 1st Hussars, in 1759. Taking advantage of the King's favour, von Kleist raised his first squadrons of dragoons and hussars in the same year, 1760. A battalion of infantry (the 'Green Croats') followed in 1761, and another in 1762. These foot-soldiers wore a hybrid uniform, composed of virtually all the typical 'hussar' impedimenta, including the *mirliton*, but with a uniform coat replacing the dolman. Attached to this picturesque force was a chasseur detachment dressed in entirely orthodox green uniforms with red facings.

Kleist's dragoons, ten squadrons strong, had the signal privilege of being allowed to carry standards exactly like those of the regular troops. His hussars also reached the strength of ten squadrons within two years, and were formed into a regiment in 1762.

There were also ten squadrons of uhlans, dressed in the Polish fashion, to add a note of exotic originality to this powerful private army, which fought mainly in Saxony and Pomerania until it was disbanded at the end of the Seven Years' War.

Schony's *Freikorps* was of Hungarian origin, a fact more obvious in his two squadrons of hussars than in his corps of grenadiers,[2] formed from renegades from the Austrian army. Their commander, von Schony, was himself a former Hungarian captain who received permission to raise his little army, a few thousand strong, in 1761.

PRUSSIA, FREIKORPS CAVALRY AND PROVISIONAL UNITS

1. Hussar of Colonel von Kleist's *Freikorps*, 1760. Variations known are: (1) *kolpak* with all-green cords, red collar and cuffs; (2) yellow belt with white loops and *fouet* and red scharawades; (3) kolpak with red *flamme* and white cords, green collar and cuffs, yellow frogging and buttons; (4) yellow belt with yellow cords and red loops, red scharawades and green sabretache with yellow edging. The only problem is one of too much choice, since all these variations come from first-rank authorities.—2. Dragoon of Colonel von Kleist's *Freikorps*, 1756–63.—2a. Alternative headgear for fig. 2.—3. Provincial regiment of Pomeranian hussars (also known as Hohendorff's) in 1757.—4. Provincial regiment of Neumark Hussars, 1757–63.—5. Provincial regiment of Kurmark Hussars, 1758–62.—6. *Freikorps* of Lubomirski's Hussars, 1758.—7. *Freikorps* of Bequignolles's Hussars (also known as Truembach's, or the White Hussars, or the Prussian Volunteers), 1760.—8. Regiment of Bawer's (or Bauer's) Hussars, also known as Pfuhl's, 1761. Another version is very similar to the uniform in fig. 1 (von Kleist's) but with a white cord on the kolpak, yellow buttons and frogging, red belt with yellow loops and green sabretache with yellow braid. It seems unlikely, however, that the influential Colonel von Kleist would have tolerated the existence of a uniform so similar to that of his own troops.—9. Favrat's corps of Hussars, 1763. The mirliton is also known with yellow cord and blue tassels.—10. Mayr's (or Coilignon's) corps of hussars, 1758.—11. Glasenapp's hussar *Freikorps*, 1760.—12. Schony's hussar *Freikorps*, also known as the Hungarian Hussars, 1761.—13. Uhlan of von Kleist's *Freikorps* between 1756 and 1763. The sabre and sabretache were worn under the caftan.

---
[1] See Vol. I, p. 152.
[2] See Vol. I, p. 151, fig. 9.

The Uhlans

The first uhlan regiment was formed in 1741 from a squadron of so-called 'Polish uhlans' recruited from the Wallachian tartars and the minor Polish aristocracy. But these first uhlans, unfamiliar with the difficult techniques of fighting with the lance, were converted to hussars the following year, as the 4th Regiment.

The uhlans reappeared in Prussia in 1745, this time in the shape of a small band of some fifty horsemen. Abandoned without money or provisions by a Saxon paymaster with a severe case of gambling fever, they deserted and offered their services to the King of Prussia. Known as 'Bosniaks', although in fact they were composed almost entirely of Cossacks, Tartars, Turks and Albanians, they were attached first to Zieten's 2nd Regiment and then to the 8th ('Death's Head') Hussars, before eventually forming a new regiment of their own, the 9th.

The illustrations show their first uniform; the second was again red, with wide, but not baggy trousers. The caftan was abandoned, as was the turban, which gave way to a hussar-style kolpak (though without a *flamme*). In winter the uhlans wore a long, voluminous dark blue frock-coat with white trimmings.

The Artillery

At the beginning of the century artillery officers wore a red coat with gold lace and facings of 'death-bed blue' (very pale blue). Under this they had waistcoats and breeches of pale yellow or straw colour; this was also the colour of the cloth trimmings on the blue coats worn by their men. White stockings were worn by officers and men alike. Frederick William I, the Soldier King, did away with the red coat, replacing it with the blue uniform with red facings then general throughout his army. The only exception to this austere uniformity was the introduction of a black mitre with a small copper peak for the bombardiers.

The Soldier King's son, Frederick II, made no change in the artillery uniform, but issued the sappers with red mitres with white metal front plates. Even the establishment of field artillery units brought no change to the uniform apart from the appearance of the essential cavalry boots.

Under Frederick William II there appeared the dark blue lapels later to be familiar under Frederick William III after 1798, in a slightly different form, buttoned all the way up and made of black cloth, like the collar and cuffs.

Headgear still followed the usual pattern, including the inelegant *Kasket*,[1] embellished with a copper grenade.

Frederick the Great may have been satisfied with the uniform he inherited from his father, but where equipment and tactics were concerned he instituted a series of far-reaching reforms which made the Prussian artillery the envy of Europe. He provided his army with copious field artillery, independent and above all highly mobile: it was to this last quality that the King devoted most of his attention and energy. He drilled his men

PRUSSIA, ARTILLERY, ENGINEERS AND AUXILIARIES
1. One of Natzmer's Uhlans, sometimes known as the Polish Squadron, in 1740. This unit became a regiment of uhlans in 1741, then was converted into Natzmer's (4th) Regiment of Hussars in 1742.—2. Artilleryman in 1709.—3. Bombardier N.C.O. in the reign of Frederick II, with the old 1731 mitre which became obsolete in 1750.—4. Artilleryman in 1760—the uniform is identical to that in fig. 3.—5. Officer of the sapper-pioneers.—6–7. Sappers.—8. Officer of the Corps of Engineers.—9. Mounted artilleryman, with the plume introduced after the Seven Years' War. An officer wore the same uniform, with the variations shown in fig. 10.—10. Artillery officer in the reign of Frederick II.—11. Mounted chasseur of the *Feldjägerkorps*. This force, 60 men strong, was raised by Frederick II at the time of his accession to ensure the safe passage of his mail, both in peace and in war. In 1744 the strength of the force was increased to eight officers and 168 men under the overall command of the King's *General-adjutant*.

[1] See Vol. I, pp. 152 and 154.

unsparingly, forcing them to achieve the coveted mobility whether they liked it or not, with spartan discipline and intensive training. These were the only answers to the confusion of the battlefield, where artillery so often lost most of its effectiveness. Frederick II's horse artillery was to be the first of its kind capable of following up and effectively supporting the attacks made by the cavalry. 'Old Fritz' left Prussia with a superbly organized army that remained unchanged until 1809.

Gerhard David Scharnhorst (1755–1813), son of a peasant, was in 1780 an artillery lieutenant in the Hanoverian service. In 1786 he wrote his *Officer's Manual*, which offered the first logical thesis on the use of artillery, together with some realistic suggestions about the employment of the horse artillery. In 1792, commanding some Hanoverian auxiliaries in the English service fighting against the French in Holland, Scharnhorst was able to expose the weaknesses of traditional tactics. By 1796 he was a major, but was refused command of a regiment for no better reason than that he had risen from the ranks. Embittered by this, he went to the King of Prussia in 1802 with an astounding request: he asked permission to reorganize the Prussian army according to his own theories, with a title and the rank of colonel as the prizes if he should succeed. More astonishing still, Frederick William III accepted the offer!

Scharnhorst was unable to prevent the catastrophe of the war of 1800 against France, but he learned from it. Now a Major-General, he realized that the day of the professional army was over; he saw the need to build a citizen army, based on conscription and fighting for patriotism, not pay; a powerful, *national* army. He dismissed incompetent or inadequate officers to forge the army of the so-called War of Liberation (1813–15). This great reforming general was mortally wounded at the Battle of Lützen[1] on 2 May 1813 which saw the first violent clash between French conscripts and German students.

---

[1] A town in eastern Germany, south-west of Leipzig, where Napoleon defeated the Russians and Prussians. It was here, too, that Gustavus Adolphus of Sweden was killed in action on 16 November 1632.
[2] It was he who exchanged his famous dragoons for Frederick William I's porcelain collection.

Saxony

For Saxony, eighteenth-century military history began punctually in 1700. In that year she signed an alliance with Denmark and Russia against Sweden. Her reward was the invasion of her own territory, and the loss of Poland – the Saxon Elector Frederick Augustus I had been crowned King of Poland as Augustus II in 1696.[2] But the military reverses which his conqueror Charles XII suffered in Russia at the hands of Peter the Great's army allowed Frederick Augustus to reoccupy his Polish throne easily enough in 1709. The new incumbent, the Pole Stanislas Leszczynski, who had been installed by the Swedes in 1704, was finally eliminated in 1733 by the new Elector Frederick Augustus II, now Augustus III of Poland. But between Saxony and her Polish kingdom there still lay the Austrian territory of Silesia.

When Frederick II of Prussia marched on Silesia in 1741, on the outbreak of the War of Austrian Succession, Saxony promptly joined forces with the invader, only to see the coveted booty handed to Prussia in 1742 by the Treaty

SAXONY (I)
1. Musketeer of von Friesen's Regiment (Seven Years' War). — 2. Musketeer of Prince Charles's Regiment, 1771. This is the new-style uniform of 1765, except for the waistcoat, which at this period was still in the regimental colour. The uniform worn between 1771 and 1810, with breeches and half-boots adorned with Hungarian knots, was otherwise the same as shown here. The infantry coat, which had been red since 1695, changed to white in 1734, except for the *Leibgrenadiergarde*, which retained the original red. — 3. Grenadier of the *Leibgrenadiergarde* in manoeuvre dress (without coat), 1791. The coat, still red, no longer had decorated buttonholes. — 4. Grenadier of the *Leibgrenadiergarde* during the Seven Years' War. — 5. Drummer of Marchen's Regiment in 1762. — 6. Artilleryman during the Seven Years' War. The green coat and red colours remained current from 1717 until 1914, except for a brief period between 1728 and 1730 when the red was replaced by pale yellow. — 7. Ensign of the Princess Palatine's Regiment of Grenadiers in 1754. — 8. Artilleryman in 1799. — 9. Artillery officer in 1799. — 10. Engineer officer in 1799. — 11. Artillery train in 1799.

of Breslau between Prussia and Austria. Undeterred, the shameless Frederick Augustus changed sides and allied with Austria in the hope of obtaining his object by a different route, but in vain.

Hostilities between Austria and Prussia flared up again with the Seven Years' War. Frederick II anticipated the Austrian attempt to reconquer Silesia by himself mounting a preemptive invasion of Saxony. Dresden fell without resistance, and the bulk of the Elector's army surrendered at Pirna[1] in 1756, delivering to Frederick 17,000 men complete with weapons, baggage and colours. In accordance with the usual chivalrous courtesies of the time, the officers of the defeated army were invited to dine with their conquerors, then given their freedom in exchange for an undertaking that they would not bear arms against Prussia again for the duration of the war.

The common soldiers were less kindly treated, being forced to swear loyalty to Frederick the Great's flag and then stuffed into Prussian uniforms. Their N.C.O.s were promoted to ensigns or lieutenants but were entirely under the thumb of the Prussian King's own officers.

The ranks of the unfortunate men who had surrendered at Pirna were swelled by 9,284 Saxon recruits; mutinies soon broke out in every major town in the country, and even on the battlefields, where the Saxons had no qualms about turning on their 'brothers in arms'. Eventually these 'reluctant Prussians' fled into Poland, or took refuge with the allies fighting against Prussia. In the following year they paid the price, along with the French, in the bloody defeat of Rossbach.

Hardly had the war ended when first the Elector then his son Frederick Christian died, leaving the throne to a boy of thirteen, Frederick Augustus III. Saxony now entered a relatively long period of peace and economic recovery, until she took part in the short War of Bavarian Succession (1778–79), and then allied with the Austrian and Prussian forces against the French revolutionary armies in 1792–96. Ten years later the Saxon armies were again in the field, this time against Napoleon, but eventually the peace of 1806 made Saxony a Kingdom, and one of the most faithful allies of the French Empire.

SAXONY (II)

1. Dragoon of Count Brühl's Regiment in 1756. The saddlecloth of his horse was the same blue as the uniform, edged with the same braid as the mitre.—2. Regiment of the *gardes du corps* in 1734. The buttonholes became rather narrower after 1741, and disappeared altogether in 1745. For active service, the coat was left off and a cuirass breastplate was worn over the waistcoat, resulting in a similar appearance to fig. 5. The saddlecloth was of the same pattern as that in fig. 5 but in sky-blue, with two rows of yellow braid, until 1756, when the Elector of Saxony's army surrendered to Frederick II. At that time the regiment's full-dress uniform included a surcoat identical to that shown in fig. 6, but with gold trimming instead of silver.—3. Officer of Prince Charles's Light Horse in 1756. Note the unusual little lapels on the waistcoat. Troopers in this unit were equipped like the dragoon in fig. 1.—4. Hussar of von Schill's partisans in 1761. In 1809 von Schill's son Ferdinand (1773–1809) raised a partisan force on his own account and invaded Westphalia, hoping to stir up a general uprising against Napoleon throughout Germany. After some initial success he and his force were crushed at Stralsund. Those officers who managed to escape were reduced to the ranks by the King of Prussia, while eleven others were shot at Napoleon's order at Wesel. One sidelight on this, as macabre as it is unusual: Jerome Bonaparte set a price of 10,000 francs on von Schill's head, which was concealed at Leyden in Holland at the house of the famous doctor and naturalist Brugmans. On Brugmans's death the carboy of wine alcohol with its sinister contents passed to the anatomical collection of Leyden University's museum, which in turn donated it in 1837 to the city of Brunswick, where the head was finally buried alongside the remains of the eleven officers.—5. Cuirassier of the Elector's Regiment in 1785. White plumes were worn from 1795 onwards.—6. *Trabant* of the Grand Musketeers in 1730; the 'troopers' in this unit, known as *Trabants*, ranked equal with lieutenants in the rest of the army.

---

[1] South of Dresden.

Bavaria

In 1701 Bavaria entered the War of Spanish Succession on the Franco-Spanish side and her armies suffered with those of her allies in the disaster of Blenheim.[1] The Elector Maximilian II Emmanuel did not regain his lost territories until 1714.

In 1741 his son Charles Albert plunged into that most chaotic of all wars, the War of Austrian Succession. Charles VI, Emperor of Austria, had just died, and the terms of the Pragmatic Sanction required that his daughter, Maria Theresa, should succeed to the throne. Most of the signatories promptly forgot their promises, and Charles Albert – who had himself signed the famous act – claimed the Austrian Empire as his inheritance.[2] Bavaria was joined by a massive coalition of Sardinia, Saxony, Spain, France and Prussia for her assault on Austria.

The Bavarian army, with powerful support from the French, took Prague and Bohemia. The Elector was crowned King of Bohemia. Shortly afterwards, at Frankfurt, he was anointed and crowned as Emperor Charles VII of Austria. This emperor without an empire enjoyed only a brief 'reign'; he was hounded out almost at once by the 'pragmatic' army of Austria with her English, Dutch, Hessian and Hanoverian allies. He died in 1745, and this strange interlude was at an end. His son Maximilian was happy enough to cast his vote, as Elector, in favour of the new Emperor Francis I of Lorraine, Maria Theresa's husband, in exchange for the return of his Bavarian lands.

In 1777 the Bavarian branch of the House of Wittelsbach became extinct with the death of Maximilian Joseph III, and Bavaria passed to the Elector Palatine, Charles Theodor. Having no direct heir himself he encouraged military intervention by Austria, who was claiming one-third of Bavaria as her own, hoping thus to secure for his illegitimate sons the rank of Princes of the Austrian Empire. This manoeuvre provoked the intervention of Frederick II of Prussia and Charles Duke of Zweibrücken, heir presumptive to the devious Elector's title: this was the brief War of Bavarian Succession, which was concluded by the peace of Teschen in 1779. Austria kept the Inn Valley region, Prussia collected Franconia as her share of the spoils, and Charles of Zweibrücken's inheritance was guaranteed.

---

[1] See Vol. I, p. 94.
[2] On the grounds that his wife, Maria Amelia, was the daughter of the Emperor Joseph, brother and predecessor of Charles VI.

BAVARIA (I)
1. Archer of the Prince Elector's palace guard in 1760, wearing the ceremonial sleeveless surtout over the uniform coat.—2. Left, fusilier of Count Charles of Preysing's Regiment with coat; right, the same in tunic only. Both in 1742.—3. Officer of the same regiment during the Seven Years' War.—4. Officer of Minucci's Regiment, 1742.—5. Grenadier of the Truchsess Regiment, 1742.—6. Fusilier of Minucci's Regiment during the Seven Years' War.—7. Fusilier of Preysing's Regiment in 1796. He wears the new, 'Rumford' uniform, an oddity of which was that the tunic was sewn to the inside of the coat, and the gaiters to the legs of the breeches.—8. Grenadier of the *Leibregiment* in 1722.—9. Musketeer of the *Leibregiment* in 1701. The Prince Elector's Regiment and the grenadiers both had braided buttonholes as a token of their elite status. For colours, see text.— 10. Grenadier of the *Leibregiment* in 'camisole' (tunic). The coat was blue with white facings and cuffs decorated in the same style as the tunic.—11. Grenadier of the *Leibregiment* during the Seven Years' War. The regimental black was introduced at the same time as the 'bastion' buttonholes. The moustache was a privilege of the grenadiers: those with fair hair were required to dye their moustaches, while those incapable of growing one at all improvised with false hair or even paint. The curled ends had to be waxed to keep them tidy.—12. The helmet known as the 'Rumford *kasket*'. The tassle was white for grenadiers and cavalry, black for fusiliers, chasseurs and garrison troops.

The spectacular successes of the French revolutionary armies against the First Coalition enabled them to take possession first of the Palatinate, then of Bavaria, between 1792 and 1795. Devastated by war, Bavaria was reoccupied by the Austrians in 1799, then retaken by the French the following year. The new Elector, Charles of Zweibrücken's younger brother Maximilian Joseph IV, found himself in 1799 inheriting a bankrupt economy and a ruined army. Becoming sympathetic to French ideas, he signed a separate peace with the conqueror in 1801, and was quick to stand beside his new ally against the forces of the Third Coalition in 1805: at Ulm and Austerlitz 30,000 Bavarians fought with Napoleon's armies. Napoleon conferred on Maximilian Joseph the title of King of Bavaria, and the new Kingdom became part of the Confederation of the Rhine; it was to survive until 1918.

BAVARIA (II)
1. N.C.O. in Prince Hohenzollern's Dragoons (Seven Years' War).—2. Cuirassier of Prince Taxis's Regiment (Seven Years' War).—3. Cuirassier of Costa's Regiment, 1742-45.—4. Mounted grenadier of the Empress's Regiment.—5. Cuirassier of Minucci's Regiment in 1792 in the 'Rumford' helmet. The cuirass was only worn up until 1785.—6. Artilleryman, 1742-45.—7. Artilleryman during the Seven Years' War.

OTHER GERMAN UNIFORMS (pp. 90–1)
1. *Württemberg*
a. Grenadier.—b. Cuirassier.—c. Grenadier. In combat the mitre was provided with a cover, to avoid confusion with an identically dressed Prussian regiment.—d. Musketeer of a garrison regiment.—e. Mounted chasseur.—f. Fusilier.—g. Horse grenadier. He wears a polished iron breastplate under his coat.
2. *Landgraviate of Hesse-Kassel (I)*
a. Cuirassier.—b. Grenadier.—c. Musketeer.
3. *Hanover*
a. Dragoon.—b. Grenadier.—c. Horse grenadier.—d. Musketeer.—e. Cavalryman in 1700.
4. *Baden*
a. Foot-soldier in 1730.—b. Ensign.—c. Grenadier.—d. Fusilier of an elite regiment.—e. Grenadier of an elite regiment.
5. *Hesse-Kassel (II)*
a. *Freikorps* musketeer.—b. Grenadier of an elite regiment.—c. Grenadier.—d. Ensign.
6. *Palatinate*
a. Officer of the mounted carabineers. He wears a polished cuirass under his coat.—b. Horse grenadier.—c. Grenadier of the Guards Regiment.
7. *Brunswick*
a. Hussar.—b. Musketeer.—c. Infantry officer.—d. Drummer of an elite regiment.
8. *Hanseatic Cities*
a. Lübeck grenadier.—b. Bremen grenadier.
9. *Saxon and Westphalian Provinces*
a. Musketeer.—b. Chasseur.—c. Grenadier.—d. Black carabineer, Schaumburg-Lippe. These survivors of the Thirty Years' War were regarded as light cavalry.
10. *Upper Rhine Provinces*
Grenadier of an elite regiment.
11. *Swabian Provinces*
a. Musketeer.—b. Grenadier.
12. *Rhenish Marches*
Grenadier.
13. *Franconian Provinces*
a. Artilleryman.—b. Dragoon.
Except where the figure is specifically dated, all uniforms shown on these two pages date from the Seven Years' War. Apart from Brunswick, Hesse-Kassel and a few minor Saxon principalities, all these troops fought against Frederick the Great. One should not be deceived by this parade of bright uniforms: the men inside them often endured the worst privations, notably those Franconian troops who fought in the army of the Holy Roman Empire, who for ten days before Rossbach had no tentage and virtually no bread or water.

AUSTRIA, BELGIUM, ITALY AND SPAIN

This part of our second volume is largely concerned with another of the Great Powers: Austria. We begin with the armies of Austria itself, infantry and cavalry; from the Austrian Walloon regiments it is a natural enough step to Belgium, which existed within the Austrian sphere of influence until breaking away at the end of the century: then come Italy, partly under Austrian occupation, and Spain, with her own interests to look after in the Italian peninsula.

Austrian Infantry: the Fusiliers

In 1700 Austria had twenty-nine infantry regiments – known as *Regimenter zu Fuss*, or literally foot regiments – each comprising three battalions of four companies, and including a regimental artillery of two or three guns.

In common with all the other armies, the Austrian one wore a uniform that differed little from civilian dress, except for its pearl-grey colour and vast, brightly coloured cuffs.

Private soldiers wore the *Halsbinde* – neckerchief or cravat – knotted at the back, corporals at the front. Companies were distinguished from one another by different-coloured rosettes worn in the hat. Ammunition pouches were stamped with the coat of arms of the Colonel in Chief of the regiment. On summer campaigns a little spray of leaves was worn in the hat, replaced in winter by a bunch of corn-stalks.

From 1718 onwards the 'foot regiments' became formally known as 'infantry'.

The Grenadiers

These elite troops affected a particularly military appearance, with their swarthy faces and bristling moustaches. They were distinguished from the other troops by their headgear, a sort of bearskin cap with a dangling streamer of brightly coloured cloth, the *flamme*, which explains the popular name of 'fused hat'. They wore two pouches, one standard one for cartridges, and another extra-large one for the grenades, which was abandoned when, around 1740, the grenade ceased to be used in major campaigns. At this period there were two grenadier companies to each regiment.

AUSTRIA, GERMAN INFANTRY (I)
1. Fusilier in 1710. – 2. Drummer in 1710. – 3. Drummer in 1769. – 3a. Wings worn by fifers and drummers in the German regiments. – 3b. Details of the drummers' and fifers' uniforms in the Hungarian regiments. – 4. Senior officer, 1740–60. – 5. Junior officer, 1740–60. – 6. Grenadier in 1710. One can see the central pouch containing the musket ammunition; the larger pouch was for the grenades. – 7. Grenadier in 1740. – 8. Grenadier in 1760. – 9. Grenadier in 1767. – 10. Grenadier in 1790.

1740–1800

In this period a distinction came to be recognized between 'heavy infantry', made up of German recruits, and the 'light infantry' recruited in Hungary. As well as these two types, frontier regiments were raised, consisting of Bosniak troops.

In 1745 the Austrian troops became known as 'imperial and royal' – *Kaiserlich-Königlich*, often abbreviated to *K.u.K.*[1] In 1742 the infantry was sixty regiments strong, and in 1756 each *K.u.K. Infanterieregiment* comprised three battalions of fusiliers. In 1769 regimental numbers were introduced, being placed before the name of the Colonel in Chief in the regiment's title.

For the Seven Years' War, white uniforms were introduced, and from 1767 onwards only a single row of buttons was worn. At this time the colours of the various regiments were laid down, to remain unchanged in some cases until the end of the First World War.

It was at this period, too, that a new type of headgear was introduced – the *Kasket*. This was a sort of bastardized mitre, made of felt, with a black leather front plate (bordered with copper until 1780) embellished with a baroque-style plaque bearing the imperial monogram. In 1790 the monogram gave way to the double-headed eagle, and the grenadier cap lost its dangling *flamme*.

By 1791 the infantry consisted of fifty-nine line regiments and seventeen *Grenzregimenter* or frontier regiments. Military service, originally for life, was reduced to ten years in 1802, though only in the 'German' regiments.

[1] Francis of Lorraine, husband of Maria Theresa, had just been elected Emperor (see chapter on 'Bavaria', p. 86), while Maria Theresa herself was Queen of Hungary. She wanted to stress the fact, to please her Hungarian subjects, whose support was essential to her (see below, section on 'The Hussars', p. 112). Hungary came to enjoy an ever-increasing measure of autonomy within the framework of the 'dual monarchy' symbolized by the name 'Austria-Hungary' by which the Habsburg domains were called until 1918.

During 1799 a new style of headgear was brought in: this was one of those 'classical' crested helmets so beloved of all contemporary armies, topped by a crest (silk for the officers, wool for the men). Very expensive and impractical, despite its visor and its small neck-guard, this innovation was ousted by the shako from 1806, though many examples survived in the infantry regiments until as late as 1809.

The Chasseurs

These specialist troops, intended to man the outposts and fight in irregular formation, were recruited from the most agile and independent young men available. Expert marksmen, the chasseurs were armed with high-quality carbines whose relatively short barrels were fitted with monstrous sword-bayonets so as not to put the chasseurs at a disadvantage in reach when it came to hand-to-hand fighting against the long rifles of the enemy infantry.

At first these *gelernten Jäger* – professional hunters – as they were rather curiously called, numbered only fifty; by 1756 they had expanded to two full companies, and by 1760 to ten. At first their duties were confined to protecting the pioneer units, but later their potential was fully exploited

AUSTRIA, GERMAN INFANTRY (II)
1. Officer in 1767.—2. Ensign in the reign of Maria Theresa (1740–80), whose coat of arms can be seen at the centre of the emblem.—3. Foot-soldier, 1767 to 1798. The identical equipment can be seen in fig. 7 of the plate showing the Hungarian infantry.—4. Chasseur of the short-lived *Feldjägerkorps*. The best Austro-German sources give the background colour as grey, but it is shown as a handsome sky-blue in the precious Vienna manuscript to which we make frequent reference, cf. especially p. 102.—5. Chasseur in 1796. A contemporary water-colour shows the same pale blue as the manuscript referred to above.—6–7. Officer and soldier, 1798.—8. Chasseur in 1798. Note the enormous bayonet, whose size was intended to compensate for the inadequate length of the rifled gun in use at the time. The 'FII' monogram on the helmet is that of Francis II, who reigned from 1792 to 1835.

in their classic skirmishing role. Drastically reduced in numbers in 1762, they were attached to the headquarters infantry regiment, and eventually disbanded in 1763. From then on the chasseurs had only a sporadic existence, as independent companies raised when the outbreak of a new war required, until the formation of the *Tiroler Jägerregiment* in 1801.

Officers and N.C.O.s

Sergeants, recognizable by the undulating blades of their halberds, also carried a baton of 'inferior wood', one inch thick and secured by a wrist-strap. Corporals also carried halberds, but with normal straight blades; their batons were virtually identical to the sergeants', but with thinner straps. All these shafted weapons were abandoned in 1769, as were those of the officers.

Austrian officers had no uniform in the strict sense before 1718, and then were not content with the pearl grey cloth worn by their men. Apart from the gold braid lace on coat and tunic, and on the pockets, officers were distinguished by a yellow and black silk sash, worn either round the waist or across the chest. In the case of general officers, this sash was made of gold cloth and black silk. Between 1743 and 1745, the period during which the crown of the Holy Roman Empire slipped from Austria's grasp,[1] the sash was green, mingled with silver and gold.

The spontoons carried by serving officers were also indicative of rank: they varied in the shape of the blade, the quality of their decoration, and the elaborateness of the fringed tassel below the blade. An officer who was not carrying a spontoon could be ranked, by an experienced eye, according to his baton, which the senior officers carried everywhere with them. An ensign's was a silver-topped cane with a thin ribbon; a lieutenant carried a substantial baton of 'Spanish bamboo' with a strap but no knob; captains and majors both had thin malacca canes, the former with a bone handle, the latter with a silver one and with a small silver chain attached. A lieutenant-colonel had a rather larger silver handle with no chain, and a full colonel qualified for a gold handle.

On occasions when spontoons had to be carried, these batons were fastened to one of the right lapel buttons.

The disappearance of the spontoon in 1769 led to the introduction of the sword-knot, which exactly reproduced the details of the fringed tassel on the spontoon, and thus served in its turn as an indication of rank.

AUSTRIA, HUNGARIAN INFANTRY
1. Grand Duke Ferdinand's Regiment, 1745. Note the sabretache, most unusual on a foot-soldier. — 2. Gyulay Regiment, 1762. — 3. Preysach Regiment, 1762. — 4. Nicolas Esterhazy Regiment, 1762. — 5. Forgach Regiment, 1762. — 6. Joseph Esterhazy Regiment, 1762.
A. The coat buttonholes shown in figs. 1–6 are arranged according to the patterns shown in contemporary sources. But one anonymous water-colour, and the Vienna manuscript, show them as much more widely spaced out. With the bottom buttonhole level with the bottom of the tunic, the coat could be fastened across the stomach in winter, as effectively as in the so-called German regiments (which had the three extra buttonholes under the coat facings found in all other armies of the period). It is thus very probable that our documents, hitherto ignored in this respect, do in fact show the correct arrangement of the buttons. For the shape of the buttonholes we have, in any case, followed the invaluable Vienna manuscript.
7. Foot-soldier in 1767. — 8. Grenadier in 1840. — 9. Grenadier in 1769. The officers wore the same uniform apart from the epaulets and leather equipment. Round the waist, an officer wore a gold scarf striped with black. The decorations on facings and breeches were gilded. He wore hussar-style boots, with gold braid and tassel. The sabre, too, was based on the hussar model, with a gold sword-knot. — 10. Grenadier in 1798. The match-holder (for the match with which he would once have lighted his grenades) had long since become purely ornamental.

---

[1] See chapter on 'Bavaria', p. 86.

The Walloon Regiments

Immediately after the victory of the Anglo-Dutch troops over the French at Ramillies in 1706,[1] large numbers of Belgians loyal to Philip v of Bourbon[2] went over to Spain. Others, sickened by the greed of the *intendants* and the crushing taxation levied by Louis XIV in the Low Countries to support his grandson's forces, enlisted in the national army raised by Charles of Habsburg in opposition to Philip. This national force consisted of seven infantry regiments – Sart, Claude de Ligne, Los Rios, Hartog, Maldeghem, Lannoy and Pancarlier – two dragoon regiments, and one of heavy cavalry.[3] They fought in the continuing wars in the Low Countries, and distinguished themselves particularly at the battles of Oudenaarde (1708) and Malplaquet (1709).[4]

The Belgians, weary of the Anglo-Batavian tyranny, welcomed the decision taken by the allies to cede their country to the Archduke Charles, who had meanwhile become Emperor of Germany. But various fortified towns had been handed over to the United Provinces, and the Belgians had to maintain the garrisons, while subject to what was virtually an Austrian occupa-

[1] See Vol. I, pp. 26 and 94.
[2] See chapter on 'Spain', pp. 124 and 126.
[3] See pp. 108 and 110.
[4] See Vol. I, pp. 26 and 94, and see also above, p. 50.

AUSTRIA, FRONTIER GUARD REGIMENTS

These regiments were recruited largely from the Serbo-Croat population and were intended to protect the empire against the Turkish menace.
1. Frontier guard in 1759.—2. Karlstädter-Oguliner Regiment.—3. Karlstädter-Ottochaner Regiment.—4. Warasdiner-Creutzer Regiment with full equipment. This soldier wore a green shoulder-pad on his left shoulder.—5. Slavonier-Brooder Regiment.—6. Karlstädter-Lykaner Regiment.—7. Karlstädter-Szluiner Regiment. The figures numbered 1 to 7 date from 1762. A seventh regiment, the Warasiner-St-Georger, was identical to fig. 4 but with dark green regimental colours and white tunic buttons. Three other regiments shown in the Vienna manuscript mentioned above have been ignored by experts of the last century, such as Knötel and Ottenfeld. Their descriptions are: (1) Banal-Grenzinfanterieregiment No. 1; as fig. 2 but with very dark blue outer garments, red tunic and cuffs, dark blue belt with red loops, and a white button with a black centre on the hat; (2) Banal-Grenzinfanterieregiment No. 2, as No. 1, but with plain hat, red belt with dark blue loops and jacket frogging replaced large red and yellow braid buttonholes grouped one, two and three as in the Hungarian infantry (see previous illustration); (3) Slavonisch-Peterwardeiner Regiment; as fig. 5 but with red instead of yellow—red belt with yellow loops, bright blue breeches, plain hat. The Viennese manuscript shows figs. 2, 6 and 7 wearing the same *opankas* as fig. 3, and also shows fig. 7 with the braid twice as close-set and even crossing over. We have on the other hand reproduced the colours shown in the original, very different from the classical impression, and also the decorations on the *klobuk* hat, hitherto much over-simplified.
8. Frontier guard in 1768.—9. Frontier guard in 1796.—10. Skirmisher of the frontier guards in 1798.

AUSTRIA, INFANTRY UNDER THE REIGN OF MARIA THERESA (pp. 100–101)

Figs. 1–8 and 13–16 are taken from Raspe's celebrated *Accurate Vorstellung der sämtlichen Kayserl. und Königl. Armee*, published at Nürnberg in 1762 and very probably based on the manuscript documents from the Vienna Albertine. This is the work used by the prodigious German authority Richard Knötel for his plates illustrating the Austrian infantry in his colossal *Uniformenkunde*.

The reader will be able to notice differences between the plates mentioned above and the diagrams on the following page, taken from a manuscript of the same year. To make comparison easier we have given not only the name of the regiment, but its number, as in the first numbering system brought into use in 1769.

The first eight figures show some of the standard manoeuvres used in loading the gun. Those that follow show some of the different ways of carrying weapons, and some of the methods of saluting taken from an illustrated book of regulations dating from the mid eighteenth century.
1. Sincere (54th), 'Cartridges ready'.—2. Kinsky (36th), 'Loading, by the left'.—3. Alt-Coloredo (20th), 'Insert ramrods'.—4. Leopold Daun (59th), 'Load'.—5. Harrach (47th), 'Replace ramrods'.—6. Baden-Durlach (27th), 'Ready'.—7. Saxe-Hildburghausen (8th), 'Aim'.—8. 'Moltke' (13th), 'Lower'.—9. Upright (officer).—10. March (officer).—11. High on the right side (officer).—12. Reversed (N.C.O.).—13. Thürheim (25th), Reversed.—14. Harsch (50th), Buried.—15. Emanuel Starhemberg (24th), Buried.—16. Mercy (56th, also called Mercy-Argenteau), 'Cover' (in case of rain).—17–20. To salute their Majesties: the halberd had to be lowered three times (fig. 18) before the movement was completed.—21–2. Royal salute for grenadier officers.—23. Saluting a general: the halberd was lowered once before adopting the attitude shown.—24. Saluting a general, for grenadier officers carrying guns; before this position was adopted the exercise shown in fig. 21 was done once.

tion. This state of affairs led to an ever-increasing discontent in the country. In addition to this, the granting to Austria of the southern areas of the Low Countries entitled her to use the Walloon troops in her own service, and they were employed against the Turks in a campaign that carried them to the very gates of Belgrade in 1717.

In 1725 the remnants of the proud regiments of 1706 were incorporated into the definitive Austrian forces in the Low Countries. The new regiments thus established were the Los Rios, Prié-Turinetti and Ligne in 1725, the Arberg in 1742, and the Vierset in 1763. Known as 'Walloon' regiments but in fact including men from all the Belgian provinces, they showed their quality in the wars of Polish Succession (1734) and Austrian Succession (1740–48), and especially at Turkheim (1745), Rocoux (1746) and Lawfeld (1747). Their exemplary courage and determination were seen again and again during the Seven Years' War, in victory and in defeat, at Prague, Kolin, Görlitz, Breslau and Leuthen; in 1758 at Hohenkirch, 1759 at Maxen, then at Landshut, then at Leignitz . . .[1]

The Treaty of Lunéville, signed on 9 February 1801, summoned the Belgians to return to their homeland. Many enlisted in the French service, though others refused. It was a Belgian escort that conducted Napoleon through Drôme on his way to Elba, and another that took the Empress and the Duke of Reichstadt to Vienna.

---

[1] We have already described several of these battles (see especially Vol. I, pp. 24, 26, 106 and 142). Rocoux (or Raucoux, now Rocourt) is a town near Liège where de Saxe defeated the Austrians and their allies – de Saxe, it will be remembered, was also the victor of Fontenoy and Lawfeld, see p. 42. As for Liegnitz (now Legnica in Polish Lower Silesia), this was where Frederick II defeated the Austrians, and where the French were also to be beaten in 1813.

AUSTRIA, INFANTRY UNIFORMS FROM 1740 TO 1780

A. Uniform from 1735 to 1767.—B. Uniform from 1767 to 1798.

All the figures showing a tricorn hat, a shoulder pad and one or the other detail are based on a Viennese manuscript with the French title *Dessins des uniformes des troupes II. et RR. de l'année 1762*. Those privileged to possess the superb *Die Osterreichische Armee von 1700 bis 1867* by Teuber and Ottenfeld, either in its original (1895) Vienna edition or in its recently reprinted limited facsimile edition will be surprised to find many discrepancies between our version of the essential elements of the 1762 uniforms and those given in the above work, which were taken from the same source. There are no less than forty-seven such differences, excluding some minor details of coat facings, arrangement of buttons on some tunics, and varying shades in the regimental colours.

1. Crown Prince of Lorraine (1726); Duke of Lorraine (1729); a: Emperor Francis I (1745); b: Emperor Joseph (1765), colour changed to Pompadour red in 1767.—2. Ujvaryi (1741); Duke Charles (1749); Ferdinand (1761).—3. Charles of Lorraine (1736), a: 1757, b: 1767; Duke Charles, 1780.—4. Hoch- und Deutschmeister.—5. Garrison Regiment 1 (1766).—6. Garrison Regiment 2 (1775).—7. Neipperg (1717), a: 1743, b: 1748, c: 1767; Harrach (1774).—8. Saxe-Hildburghausen, colours changed to red after 1732.—9. Los Rios from 1725; Clerfayt (1775).—10. Brunswick-Wolfenbüttel (1740), a: 1757, b: 1767.—11. de Wallis (François) (1739), a: 1743, b: 1748, c: 1767; de Wallis (Michel) in 1774.—12. Botta d'Adorno (1739), a: 1757, b: 1767; Khevenhüller-Metsch (1775).—13. Moltke (1737), a: 1740, b: 1743, c: 1767; Zettwitz (1780). 14. Salm-Salm (1733), a: 1740, b: 1748; Ferrari (1770); Tillier (1775).—15. Pallavicini (1736), a: 1740, b: 1757, c: 1778; Fabris after 1773.—16. Livingstein (1722), a: 1740, b: 1767.—17. Kollowrat-Krakowski (1737), a: 1740, b: 1767, c: 1770; Koch (1773).—18. Seckendorff (1719); Marshal of Bilberstein (1742); Brinken (1773). a: 1740, b: 1767.—19. Palffy (1734), a: 1740, b: 1767; Alton (1773).—20. Diesbach (1719); Colloredo-Waldsee (1744). a: 1740, b: 1743, c: 1757, d: 1760, e: 1767.—21. Schulenburg (1734); Arenberg (1754); Gemmingen (1778): a: 1740, b: 1767.—22. Suckow (1734); Roth (1741); Hagenbach (1748); Sprecher von Bernegg (1756); Lacy (1758). a: 1740, b. 1767.—23. Baden-Baden (1707), a: 1740, b: 1767; Ried (1771); Duke Ferdinand (1779).—24. Starhemberg, M. A. (1703); Starhemberg, E. M. (1741); Preiss (1771). a: 1740, b: 1767.—25. Wachtendonk (1731); Piccolomini (1741). a: 1740, b: 1742, c: 1751, d: 1767.—26. Grünne (1737); Puebla (1751); Riese (1776). a: 1740, b: 1743, c: 1767.—27. Hesse-Kassel (1732); Baden-Durlach (1753). a: 1740, b: 1767.—28. Arenberg (1716); Scerzen (1754); Wied-Runkel (1754). a: 1740, b: 1748, c: 1767, d: 1779 (Wartensleben).—29. Brunswick-Wolfenbüttel (1736); Loudon (1760). a: 1740, b: 1748, c: 1767.—30. Prié-Turinetti (1725); Saxe-Gotha (1753); Ligne (1771). a: 1740, b: 1767.—31. Haller von Hallerstein (1741), a: 1741, b: 1767; Esterhazy (1777).—32. Forgats (1741), a: 1741, b: 1767; Giulay (1773).—33. Andrassy (1744); Esterhazy, N. (153). a: 1748, b: 1762.—34. Kökemesdy de

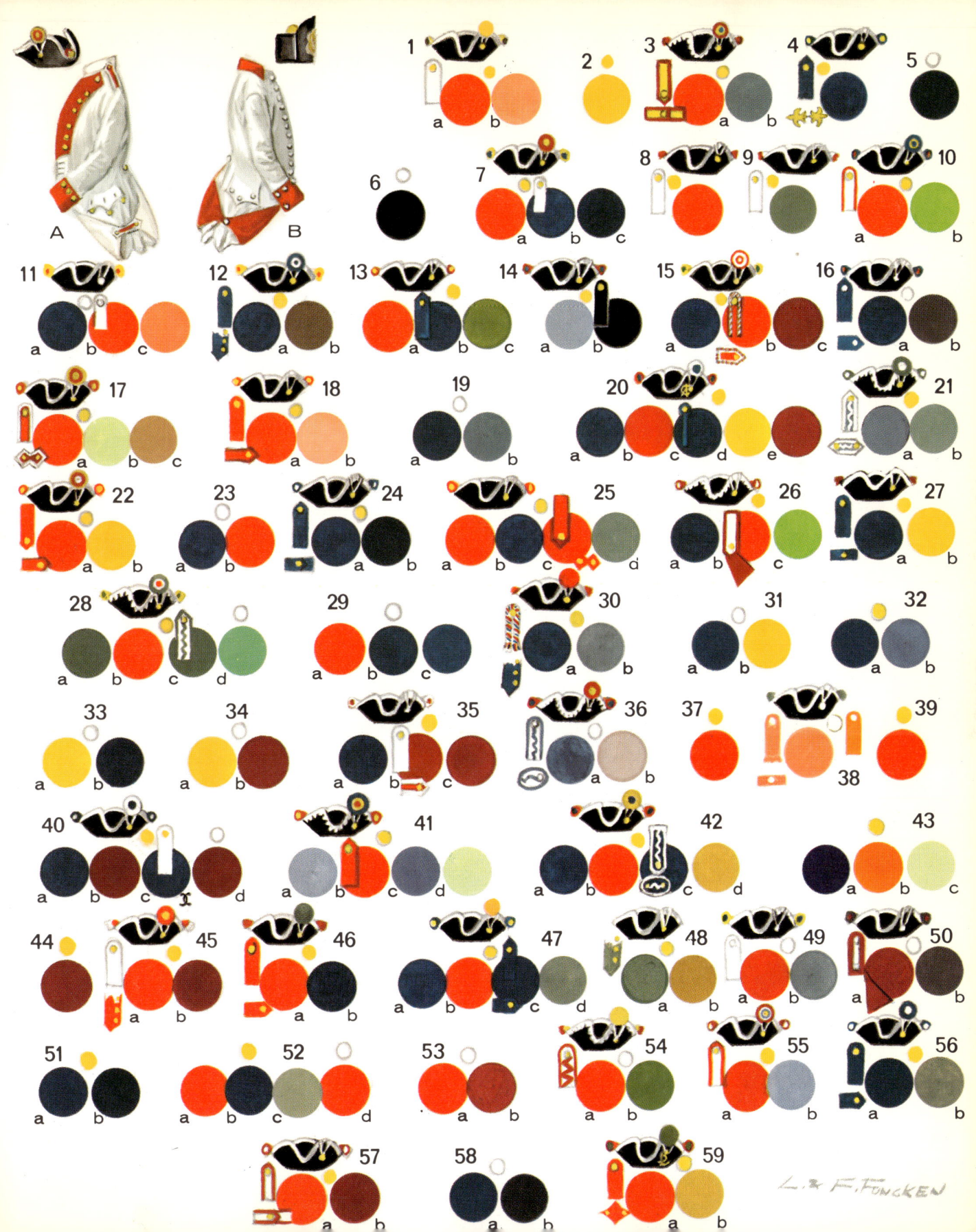

Vetès (1734); Bathyany (1756); Esterhazy, A. (1780). a: 1748, b: 1767.—35. Waldeck (1739); Marquire (1763); Hessen-Darmstadt (1767); Wallis (1774). a: 1738, b: 1743, c: 1767.—36. Browne, U. (1737); Browne, J. (1757); Tillier (1759); Kinsky (1761). a: 1738, b: 1767.—37. Szirmay (1741); Esterhazy, J. (1744); Siskovics (1762).—38. Ligne (1725); Deynse (1766); Kaunitz (1774).—39. Palffy (1756); Preysach (1758).—40. Damnitz (1734); Colloredo, C. (1754). a: 1740, b: 1743, c: 1748, d: 1767.—41. Bayreuth (1734); Plunquet (1763); Fürstenberg, C. E. (1770); Belgiojoso (1777); Bender (1778). a: 1740, b: 1743, c: 1767, d: 1770.—42. O'Nelly (1734); Gaisruck (1743); Gemmingen (1769); Mathesen (1775). a: 1740, b: 1743, c: 1757, d: 1767. —43. Platz (1737); Buttler (1767); Thurn (1775). a: 1740, b: 1743, c: 1767.—44. Clerici (1744); Gaisruck (1769); Belgiojoso (1779).—45. Daun, H. (1711); O'Kelly (1761); Bülow (1767); Lattermann (1776). a: 1740, b: 1767.—46. Spauer (1745); Ogiloy (1748); Sincere (1751); Marquire (1752); Migazzy (1764). a: 1745, b: 1767.—47. Harrach, J. (1704); margrave of Bayreuth (1764); Elrichshausen (1769); Kinsky (1779). a: 1740, b: 1743, c: 1757, d: 1767.—48. Vasquez of Binas (1734); Luzan (1755); Ried (1765); Caprara (1773). a: 1740, b: 1767.—49. Walsegg (1724); Bärnklau (1743); Kheul (1747); Angern (1758); Pellegrini (1767).: 1740, b: 1767.—50. Wurmbrand (1727); Harsch (1749); Poniatowski (1766); Stain (1773). a: 1740, b: 1767. 51. Gyulay, S. (1735); Gyulay, F. (1759). a: 1757, b: 1767.—Bethlen (1741); Karoly (1763). a: 1743, b: 1748, c: 1757, d: 1767.—53. Simbschen (1756); Beck (1763); Palffy (1768). a: 1756, b: 1767.—54. Königsegg-Rothenfels (1720); Sincere (1751); Callenberg (1769). a: 1740, b: 1767.—55. Arberg (1742), a: 1743, b: 1767; Murray (1768).—56. Daun, P. (1690); Merci-Argenteau (1741); Nugent (1767). a: 1741, b: 1767.—57. Thüngen (1735); Andlau (1745); Colloredo-Waldsee (1769). a: 1740, b: 1767.—58. Vierset (1763). a: 1763, b: 1767.—59. Daun L. J. (1740); Daun, F. (1766); Langlois (1771). a: 1741; b: 1767. (Based on H. Knötel, *Handbuch der Uniformkunde*.)

The Vienna manuscript gives some more interesting details which have not been brought out: the 26th Regiment had no buttons on its cuffs and only a single row on the tunic, a peculiarity shared with the tunics of the 3rd, 10th, 22nd, 28th, 38th, 50th and 55th. The 38th Regiment was also distinguished by a fringed epaulet on the right shoulder as well as the normal shoulder flap on the left.

Though the majority of the regiments were native Austrians, there were several exceptions: the 9th, 30th, 38th and 58th were from the Low Counties (present-day Belgium) and were called Walloon regiments, though there were many Flemings serving in them as well. The 44th and 48th were 'Italian' regiments; and the 19th, 31st, 32nd, 33rd, 34th, 37th, 39th, 51st and 52nd were Hungarian.

AUSTRIA, DRAGOONS

The figures numbered 1 to 12 show the dragoon regiments in 1762. The general appearance of a dragoon at this period is shown in figs. 13, 14 and 15, while figs. 14 and 16 to 19 show the evolution of a Walloon (i.e., Belgian) dragoon regiment, one of the most illustrious of all Austrian cavalry regiments. 1. Archduke Joseph's Regiment.—2. Kolowrat-Krakowski Regiment.—3. Duke of Württemberg's Regiment.—4. Prince Eugene of Savoy's Regiment.—5. Bathyany Regiment.—6. Löwenstein Regiment.—7. Saxe-Gotha Regiment.—8. Young-Modena Regiment.—9. Liechtenstein Regiment.—10. Hessen-Darmstadt Regiment.—11. Zweibrücken Regiment.—12. D'Althann Regiment.—13. *Stabsdragoner* or Headquarters Regiment in 1762.—14. Saint-Ignon's Regiment (formerly Prince Ferdinand de Ligne's) in 1762; this regiment became Count Nicolas d'Arberg's in 1779. 15. Grenadier of the Liechtenstein Dragoons in 1762.—16. Grenadier of the Count of Merode-Westerloo's Walloon Regiment of Dragoons in full dress, around 1730. In ordinary conditions the grenadiers wore the tricorn hat used by the other companies, and could then be distinguished only by the curve of their swords. The regiment became the property of the Prince of Ligne in 1732 (see fig. 17).—17. Walloon Dragoon of Prince Ferdinand of Ligne's Regiment in 1757. In 1759 the regiment was taken over by the Count of Saint-Ignon (see fig. 14). These were the Prince of Ligne's 'baby-faces' who sealed the victory of the Imperial forces over Frederick II at Kolin in Bavaria on 18 June 1757.—18. Latour's Dragoons, 1792. By special dispensation this regiment was allowed to change its white uniform (see fig. 19) for the traditional green coat. The shabraque was identical to that shown in fig. 17, but with the crowned monogram F.I. The colonel in chief was Fieldmarshal-lieutenant Count Maximilian of Baillet-Latour.—19. Dragoon of Count Nicolas d'Arberg's Regiment (formerly Saint-Ignon's) in white uniform, 1780. The collar and cuffs were light blue and the shabraque the same as that shown in fig. 17 but with the crowned monogram J$_{II}$ (Joseph II).

Cavalry: the Dragoons

At the beginning of the eighteenth century the eleven Austrian dragoon regiments differed from their comrades in the cuirassiers only in wearing the tricorn hat and not using the breastplate. The colours of their uniform coats varied greatly from one regiment to another.

Originally, like other dragoons elsewhere, the Austrian variety fought both on foot and as cavalry: they were, it was said, *halb Mensch, halb Vieh* ('half man, half beast') or even *weder Mensch noch Vieh* ('neither man nor beast'). As for the origins of the term 'dragoon', German historians have fared no better at unravelling them than have their colleagues elsewhere.

These mounted infantrymen (*aufs Pferd gesetzte Infanterie*) showed their origins clearly enough both in the cut of their uniforms and in their use of drums instead of the traditional cavalry trumpets. And this relationship to the regular infantry was further emphasized in 1715 by the creation of *Eliteabteilungen* of grenadiers, dressed in the same bearskin hats as their infantry equivalents. Towards the end of the Seven Years' War these horse grenadiers were equipped with a remarkable and unique weapon – the so-called *Granatpistole* or *Granatmörser*, a pistol capable of firing three-pound grenades.

It was during this same war that Austria first felt the need for the introduction of light cavalry. The first regiment of *Chevaulégers* was created from the Prince of Löwenstein's Dragoons (Fig. 6) who thus became known from 1759 onwards as the 'Chevaulegersregiment Christian Fürst zu Löwenstein-Wertheim'. Between 1760 and 1765 the same transformation overtook the Duke of Württemberg's, Saxe-Gotha, Young-Modena, Zweibrücken and Archduke Joseph's regiments in turn (Figs. 3, 7, 8, 11 and 1 respectively). (It should be mentioned here that the coloured breeches shown in our illustrations are depicted by one contemporary source as being uniformly buff.) After the end of the Seven Years' War two-thirds of the light cavalry regiments were reabsorbed into the regular dragoons, leaving only two regiments of *Chevaulégers*. In 1767 they wore a white or green uniform and the *Kasket*, whereas the dragoons always wore white, and retained the tricorn. In 1798 the two were amalgamated as 'light dragoons', but this arrangement lasted only until 1801.

AUSTRIA, DRAGOONS, LIGHT HORSE AND FREIKORPS

1. Mahony's Serbian *Freikorps*.—2. Odonel's Serbian *Freikorps*.—3. Officer of Wurmser's *Freikorps* in 1793. This unit, raised in Slavonia, fought the French in Alsace. Note the ramrod worn on the chest, for loading the pistols.—4. Viennese volunteer in 1795, wearing the so-called Corsican hat. The six corps of Viennese volunteers, 37,000 men in all, were raised to counter the threat of revolution, and showed more enthusiasm than real effectiveness. Shortly after this time, the big tricorn hat of classic tradition was extended to the foot-soldiers.—5. Mounted Viennese volunteer in 1795. Two years later, infantry and cavalry were equipped with an identical uniform, but in pale green; the cavalry also had overbreeches of the same colour which buttoned down the whole length of red braid, running down the outside of the legs. (See the Austrian Hussars, and the last figure in the last illustration of the Prussian Hussars).—6. Light Dragoon. From 1798 to 1801 both light horse and dragoons were known as light dragoons.—7. Trooper of the light horse in 1767. After the Seven Years' War, the light horse could be distinguished from the dragoons by their use of the helmet called the *Kasket*. The coat might be green or white, depending on the regiment. The horses' trappings were as shown in fig. 17 on the previous page, but with the royal monogram.

The Belgian Dragoons

Three Belgian national cavalry regiments had fought in the Low Countries war up until 1713: the Westerloo Cuirassiers and the Ligne and Holstein-Norburg Dragoons. In 1725, following the events already outlined in the section on the Walloon infantry, the remnants of these units were formed into a cavalry regiment. These survivors numbered sixty-two officers and 1,117 men. They formed a dragoon regiment under the command of a Field Marshal, the Count of Mérode, Marquis of Westerloo. On his death in 1732 these Mérode-Westerloo Dragoons became the property of Field-Marshal Prince Ferdinand of Ligne. Now known as Ligne's Dragoons, they were to play a vital role in the victory of the Imperial troops over Frederick the Great at Kolin in 1757, the year of their new commander's death.

When a Prussian victory appeared inevitable, their colonel, François Florent Count of Thiennes, repeatedly demanded permission from his commander in chief to lead a charge, furious at 'being marched 300 leagues to be –––ed about'.

'You won't get far with that baby-faced lot,' was the reply (a reference to the beardless faces of many of the young dragoons).

'We'll see about that,' retorted Thiennes; and having passed on his superior's opinion to his men, he added, 'Right, let's show them that you don't need beards so long as you've got teeth!'

Charging at full gallop, and rallying several allied cavalry squadrons in their support, Ligne's dragoons flung themselves on the spearhead of Frederick's infantry, broke through, wheeled, broke through again and again, supported each time by increasing numbers of the Imperial cavalry. The grenadiers of the Prussian Guard were thrown into the fight, but in vain; their sacrifice met with no more success than those of any of the other battalions mown down by the hurricane. The battle became a complete rout. One may well doubt the authenticity of Thiennes's 'historic' exchange with his superior, but there is no denying the concrete facts of the honours heaped on the Ligne Regiment after this great day. The Empress Maria Theresa conferred on the Belgian dragoons the privilege of never wearing moustaches – 'Baby faces they were and will remain'[1] – and presented them with four standards 'embroidered with her own hands' ... assisted no doubt by the hands of the other ladies of the court.

In 1757 the command of the regiment was given to Count Daun, who exchanged it a few months later for the Prince of Löwenstein's cuirassiers. In 1759 the Count of Saint-Ignon assumed command; and the following year, at Torgau,[2] the regiment paid the price of its brilliant success at Kolin. After a clash with Zieten's famous hussars, the Belgians were themselves massacred almost to a man by the Prussian cavalry. The survivors paid their own ransoms, and the regiment was rapidly reformed, to avenge its humiliation by overwhelming a Prussian force near Landshut[3] in 1761. In 1779 the command passed to Count Nicolas d'Arberg, and then to the Count of Baillet-Latour. Bound in honour by the

AUSTRIA, CUIRASSIERS (I)
A. Cuirassier in 1700. With his *Eisenhaube* he still resembles a cuirassier of the Thirty Years' War. His heavy straight sword, the *Pallasch*, is similar to those in the following illustrations. – B. Cuirassier in 1722.
The cuirassier regiments in 1762: 1. 1st (Stampach's) Regiment. The coat had the same arrangement of buttonholes on the left hand side, also. – 2. 2nd (Archduke Maximilan's) Regiment. – 3. 3rd (Prince Albert of Saxony's) Regiment. – 4. 4th (Bretlach) Regiment. – 5. 5th (di Stampa's) Regiment. – 6. 6th (Trauttmannsdorf) Regiment. – 7. 7th (de Ville's) Regiment. – 8. 8th (Buccow's) Regiment. – 9. 9th (Anhalt-Zerbst) Regiment. – 10. 10th (Prince Emmanuel of Portugal's) Regiment. – 11. 11th (Prince of Ansbach's) Regiment. 12. 12th (Modena) Regiment. – 13. 13th (Serbelloni's) Regiment. – 14. 14th (O'Donell's) Regiment. – 15. 15th (Palffy's) Regiment. – 16. 16th (Schmerzing's) Regiment. – 17. 17th (Benedict Daun's) Regiment. – 18. 18th (Archduke Leopold's) Regiment. – It should be noted that in combat all these regiments wore the blackened cuirass (breastplate only), secured by two leather straps crossing at the back (see the Prussian cuirassiers).

---
[1] This privilege was retained until 1918 in the 14th Austrian Dragoons.
[2] Town in eastern Germany, on the Elbe; see above, p. 66.
[3] Town in Bavaria, on the Isar.

oath they had sworn to the Emperor, Latour's Dragoons fought the Belgian revolutionaries of 1790 to the last gasp. For this faithful service the indestructible cavalrymen were given a privilege unique in the whole Austrian army – the right to enter the courtyard of the Imperial Palace in Vienna with drawn swords.

The reorganization of 1798 saw Latour's Dragoons converted into the 11th Regiment of Light Dragoons; posted to Moravia, they became the 4th *Chevaulégers* Regiment in 1802, and in 1806 came under the command of Baron Vincent.

The Cuirassiers

The cuirassier, the epitome of the heavy cavalryman, took his name from the thick leather doublet which had eventually been replaced by the detachable body-armour called the cuirass. The cuirass consisted of two parts, breastplate and backplate and was made of polished iron until 1720, after which it was blackened. Later, around 1740, the backplate was abandoned, and the breastplate was secured by straps crossing at the back. These straps were reinforced from the shoulder to the upper chest by small chains, the shape and structure of which indicated the wearer's rank. A further guide to rank was the number of studs set around the edge of the cuirass to hold the lining in place, while general officers had the entire periphery of the cuirass trimmed with gold. These elaborate distinctions were simplified from 1754 onwards.

In 1715 the eighteen regiments of cuirassiers which had existed since 1688 were each allocated a company of carabineers, an elite unit within each regiment very similar to the grenadiers in the infantry or dragoon regiments. The various carabineer companies were often brought together to form a single striking force of up to 10,000 men. In 1768 they were combined with the grenadier companies from the dragoon regiments to form two new carabineer regiments; in 1798 they were converted to ordinary cuirassiers.

The heavy and antiquated helmet, with nasal, had been abandoned very early in the century, in favour of a hat reinforced on the inside with a metal skull-cap.[1] The heavy cavalry sword used by the cuirassiers, called a *pallasch*, was nearly three feet long and, with its metal-bound leather scabbard, weighed three and a half pounds. The rest of their equipment consisted of a 17-mm carbine weighing some eight pounds and a pair of pistols, each over eighteen inches long and weighing over four pounds. This arsenal was usually employed at the halt: when the enemy attacked, the cuirassiers would begin with a volley from their carbines, followed at closer range by pistol fire, and eventually the *pallasch* would be drawn for hand-to-hand fighting.

Cuirassiers were not easy to identify by regiment, as the breastplate concealed the tunic, the colour and buttoning arrangement of which were the main regimental insignia. Until 1755 one could at least see the coat of arms of the colonel in chief, but when these were replaced on saddle-cloth and holster-covers by the double-headed eagle, only an expert eye could tell one regiment from another. In 1767 the eagle was in turn replaced by the royal monogram.

In that same year the coat, or *Koller*, now with only a single row of buttons, took on a noticeably more 'modern' appearance which it retained until 1798. At this time there also appeared an entirely new form of headgear, the 'classical'

AUSTRIAN, CUIRASSIERS (II)
1. Officer in 1740.—2. Cuirassier in 1760. The pistol was always fired in this way, tilted over to the left, to ensure the best possible contact between the powder and the touch-hole of the weapon. (See the explanation of the workings of a flintlock on p. 118 of Vol. I).—3. Cuirassier of the carabineer squadron in 1760 armed with the blunderbuss which delivered a lethal salvo of twelve balls. Note the curved sabre peculiar to the members of this special squadron.—4. Officer in 1796.—5. Cuirassier in 1770.—6. Trumpeter in 1770.—7. Ensign in 1770.—8. Cuirassier in 1798, wearing the helmet introduced that year. Shabraque and helmet both bear the monogram of the Emperor Francis II. The officer's general appearance was the same, but with a copper crest on the helmet, gilded red leather cuirass straps, and a black sheepskin over the saddle and holsters.

---

[1] This was the universal practice, though sometimes the metal protection was on the outside of the hat.

helmet, a reflection of the classicism fashionable at the time. In between times the hat had been improved by the addition of a yellow and black plume, a novelty adopted throughout the entire Austrian cavalry.

The Hussars

The first two hussar regiments in the Austrian army trace their origins back to the last two decades of the seventeenth century. Everything about them revealed their Hungarian background: the short tunic with its horizontal strips of braid, the sash, the tight breeches, the short boots known as *cismas*, and most of all the pouch called the *Säbeltasche* or sabretache, whose origins are hidden in the mists of time.[1]

As well as a curved sabre and saddle pistols, the Hungarian hussar carried a long straight sword which he used like a lance, couching it on his knee. This strange weapon was known as a *hegyestor*, and was simply left, like a skewer, in the body of the unfortunate enemy – it was then considered to have done its job and would only be recovered later if circumstances happened to permit.

Wherever they served, the hussars were at first thought of as irregulars, suitable only for skirmishing – that is, harassing the enemy and looting his baggage-train – while the regular troops fought carefully by the 'book' of planned strategy with its regular marches, its comfortable encampments and its good honest battles in broad daylight, a sort of real-life wargame whose rules no general would have considered bending.

In 1734 there existed three regiments:

	CAP LINING	DOLMAN AND CLOAK	SHABRAQUE
Karoly	Red	Light blue	Light blue
Czungenberg	Red	Green	Green
Deszöffy, Stephan	Dark blue	Dark blue	Red

AUSTRIA, HUSSARS
The regiments in 1762: 1. Emperor's Regiment.—2. Jazygier-Kumanier or Palatinal Regiment (some authorities show this regiment in crimson instead of red, with a bright blue sabretache bordered with crimson).—3. Hadik Regiment.—4. Kalnocky Regiment.—5. Baranyay Regiment.—6. Bethlen (formerly Moroth, 1741-54) Regiment.—7. Palffy Regiment (formerly Karoly, 1734-59).—8. Paul Anton Esterhazy's Regiment (also described as all blue, with blue sabretache).—9. Nadasdy Regiment (also known with red dolman, dark blue pelisse and breeches).—10. Karlstädter Regiment.—11. Warasdiner Regiment.—12. Banater Regiment (another version shows the *flamme* of the kolpak, dolman and pelisse all dark green, with buff-coloured breeches).—13. Szeczeny Regiment (formerly Festetics, 1702-42).—14. Dessöffy Regiment.—15. Splenyi Regiment (became Imre Esterhazy's during the same year, and was disbanded in 1768; a variant is known with bright blue uniform, red belt with white loops and cords, and red and white edging even round the outside of the boots).—16. Esclavonier Regiment (also shown with green cuffs and green and white braid).—17. Hussar in 1700.—18. One of Baron de Trenck's pandours in 1756. Very similar to the early hussars (see fig. 17), the pandours take their name from a small township in the Hungarian comitatus of Perth, or possibly from a slavonic tribe in the mountains of Shol. These ferocious irregular troops chose their own chief, the 'haroum-bascha' or pasha. Our illustration is the traditional impression, but Trenck himself said that he dressed his pandours in the most outlandish and lurid splendour he could manage to inspire the maximum terror in the enemy. Their armament consisted of a sabre, a musket and four pistols.—19. Hussar from 1745 to 1769.—20. Hussar from 1770 to 1798. A very short carbine was introduced in about 1790. After 1770, the frogging was black and yellow for all units. The plume, which was identical to the one in fig. 21, is here shown enclosed in its protective case of waxed cloth.—21. Hussar in 1798. The carbine is on his right side, slung from a snap-link at the end of his two crossbelts; the other held his cartridge pouch and the iron ramrod for loading his weapons. The wearing of overbreeches became standard practice during the last years of the century.

[1] Sabretache covers have been found dating back to the ninth century – this sort of wallet made up for the absence of pockets.

Braid and breeches were red in all cases until 1748, after which breeches were in natural leather.

It may seem surprising that there were so few regiments of hussars in this, their country of origin; but it must be remembered that the 'relief from the Turkish yoke' brought to Hungary by the victorious Austrian campaigns of 1683-89 had merely replaced one occupying power with another, even more rapacious. The liberators' chronic lack of cash led to the economic exploitation and systematic 'germanization' of Hungary, combined with the depredations of the 'German' troops in the garrison towns and the revival of religious persecution directed against the protestants. A revolt broke out in 1703, incited by an exiled Prince, Francis II Rakoczy, who rallied to his cause a good many aristocrats such as the Forgachs and the Esterhazys. The powerful insurgent army enjoyed some early successes and went so far as to threaten Vienna itself, but was totally and permanently crushed in 1711.

Maria Theresa's pathetic appeal to her 'loyal Hungarian subjects' at the outbreak of the War of Austrian Succession, and the proud response of the magnates of the Hungarian Diet, have passed into legend. In fact the cry of *Moriamur pro rege nostro Maria-Theresia!* ('Let us die for our "king" Maria Theresa!'), backed by the promise of an army of 100,000 men resulted in 20,000 undisciplined recruits, untrained and without equipment. The partial reconquest of Austria and Bohemia had more to do with the young empress's own determination than with the efforts of her threadbare Hungarian subjects.

Not until 1742 did the hussar regiments begin to increase in numbers, reaching a total of sixteen towards the end of the Seven Years' War.

The fur cap called the *kolpak* replaced the *mirliton* in 1767 so far as the men were concerned, the officers hanging on until 1771. The occasional modifications that took place in later years can be seen in the illustrations.

The Artillery

The excellent Austrian artillery was not organized on a permanent basis until 1756. It was divided between 'German' artillery – three brigades of eight companies, increased to ten in 1760 and Low Countries artillery – a single brigade of eight companies stationed at Malines. Increased to twelve companies after the Seven Years' War, this brigade was absorbed by the last of the three artillery regiments which emerged from the reorganization of 1772.

The cannon accompanying the infantry forces were of 75-mm calibre, as opposed to 90mm for the 'line' field artillery, 120 for the largest cannon and 150 for the mortar. The total number of cannons in the field was over 500. A bombardier corps was raised in 1786.

AUSTRIA, UHLANS AND NOBLE GUARD
1. Uhlan in 1784. In 1785 the typically Polish tunic-coat called the *kurtka* changed colour to white, with dark red trimmings, and remained so until 1792. The first Austrian uhlan regiment was not raised until 1784. After a brief interval during which it was attached to the light horse, it again became officially an uhlan regiment in 1791, numbered the 1st. A second regiment was raised during the following year, after the annexation of Galicia.—2. Uhlan in 1796.—3. Uhlan in 1798.—4. Member of the Galician noble guard in 1790. This force, also called the Polish Guard, was started in 1782 and ceased to exist in 1790. The guardsman in our picture wears a kind of cloak rolled up on his chest and thrown back over the shoulders.—5. Member of the Hungarian noble guard at the end of the eighteenth century. The Hungarian Guard of the imperial and royal court was raised in 1760 and survived until 1918. The pelisse is here worn over the dolman.

The Engineers

A single company of military engineers made its appearance in 1718. In 1747 this was expanded into an engineer corps, with a corps of sappers and another of pioneers under its command. The pontoneers, after early experiments with contraptions of leather, oiled cloth and wood, began to use sheet-metal pontoon bridges from 1735; in 1749 they were organized as a standing division of two companies, which in 1767 became a *Pontonier-bataillon* of five companies.

The pioneers with their picks and shovels had at first been attached to the artillery for navvying duties. In 1756 they formed their own batallions which disappeared in 1761, reappeared in 1778, and eventually disappeared again in 1779.

The Low Counties brigade from Belgium had been integrated with the Austrian engineers in 1770. These Belgian engineers often served with great distinction in Austria. It was one of them who was accorded the great honour of being asked to design the lavish Hofburg palace in Vienna and the triumphal arch in Innsbruck.

Belgium: the Patriots

No student of the military history of Belgium, little known as that history is, could fail to be impressed by the reputation for courage gained by the Belgians on every battlefield in Europe. As we have seen in our chapter on Austria, and will again in the chapter on Spain, there is no gainsaying the accuracy of this reputation – yet it is still completely forgotten today, especially in Belgium itself. The explanation of this phenomenon is to be found in the total silence preserved on the subject by the majority of Belgian historians during the first half of the nineteenth century.

Spain lost her 'Belgian provinces' after the Battle of Ramillies in 1706, and the territory passed under the control of Austria,[1] which administered them peacefully until the eve of the French Revolution. This was the moment picked by the Emperor Joseph II, that most disconcerting of all philosopher-kings, to hurl himself enthusiastically into a series of civil and religious reforms which provoked a major uprising. In 1787 the Emperor recalled the archdukes who governed the provinces, replacing them with the Commander in Chief of the Imperial forces in the Low Countries, Count Murray, who barely survived the outburst of popular fury occasioned by his appointment. This somewhat soft-centred soldier was next replaced by the Count of Trauttmannsdorf, who in turn relied upon the new military commander, the Count of Alton, a brutal and clumsy soldier who resorted to force, spilling blood in the course of a series of affrays in Malines, Antwerp, Louvain and Brussels during 1788.

AUSTRIA, ARTILLERY AND VARIOUS
1–3. Generals in 1720 (1) and 1760 (2–3). Note that the traditional black and gold sash (black and yellow for junior officers) was for a brief time changed to green and gold or green and silver during the reign of Charles Albert of Bavaria, crowned as German Emperor at Frankfurt in 1742 and hounded out of Bavaria by Maria Theresa during the War of Austrian Succession, 1740–8.—4. Soldier of the *Stabsregiment*, a regiment guarding the headquarters, in 1762. —5. Artillery N.C.O. in 1710. The firing trident had already lost its original function, but survived as both a weapon and a badge of rank.
6. Walloon (i.e. Belgian) artillery in 1762.—7. German artillery in 1762.—8. Fusilier of the artillery in 1762. The gun peculiar to this unit was 1.34 metres long with a calibre of 15.1mm, compared to the infantry weapon at 1.51 metres and 18.3mm. It was compulsory for an artilleryman to be able to read, write and count.—9. Sapper in 1762.—10. Pioneer in 1762.—11. Artilleryman in 1798. An officer's uniform was the same but with a taller, gilded helmet crest. The yellow and black striped belt was of course worn, and light boots replaced the gaiters shown here.—12. Sapper in 1800. He wore a short-tailed coat and a yellow cockade with a black centre at the base of his hat plume. The miner's uniform was identical, but with an all-black plume. Officers wore a similar uniform, with the usual sash of their rank and a hat trimmed with gold braid.

---

[1] See chapter on 'The Walloon Regiments', p. 98 *et seq.*

The Belgian outlaws, as they now were, were carried away by revolutionary ideas, and inspired by the events then taking place in Paris, as symbolized by the storming of the Bastille. They formed a volunteer army of 'patriots' under the command of Colonel Jean-André van der Meersch, a former officer of the 67th Regiment (La Marck's) in the French service.

Against this 'fairy army', these 'poor wretches' assembled at Turnhout in 1789, the arrogant Alton sent a force of 3,000 men. No sooner had they entered the town than the Austrians were met by a hail of fire, and fled in panic, abandoning three cannon to the enemy. This skirmish, considerably exaggerated, soon became common knowledge everywhere. Taking heart, the rebels stormed Ghent and captured its garrison of 800 men. First Diest, in Campine, and then Mons were abandoned by the Austrians and rose in revolt; finally Brussels followed suit. Seeing the cream of his Walloon regiments decimated by desertion – the Ligne, Württemberg, Clerfayt, Murray and Vierset Regiments were each reduced to a single battalion[1] – Alton abandoned the capital to seek shelter under the guns of the fortress of Luxembourg.

In January 1790 the sovereign Congress of the Republic of the United States of Belgium voted the formation of a national army of 41,498 men. By September 20,567 men and 3,645 horses had been mustered. They were led for the most part by foreign officers, lamentably incompetent apart from the Frenchman de Villers-Masbourg, the Prussian von Schönfeld and the Englishman Koehler (this last a dashing commander of the artillery). Van der Meersch, meanwhile, had been interned at Antwerp for having stood up to the new revolutionary government.

Enthusiastic but undisciplined, the makeshift army rejected Schönfeld's attempt to subject it to Prussian discipline. The Congress of Namur published a military code of remarkable ferocity which proposed to stamp out absenteeism with the flat of the sword-blade, the whip, the branding-iron and if necessary death. This intolerable project was strangled at birth by the outcry which greeted it.

Meanwhile the Imperial forces had regrouped and had pushed back the insurgents along the line of the Meuse. Under pressure from public opinion, the Belgians decided to launch a full-scale assault on 24 May. The result was a crushing setback. The spirited recruits flung themselves to the ground at the first rumble of the Austrian cannon, or turned and fled before the first wave of dragoons. The army withdrew behind a strong line of prepared fortifications. While Schönfeld, disgusted with his charges, passed the time by organizing parties for the nobility and the female canons of Namur, Koehler on the Belgian left had patiently undertaken the training of his 7,000 part-time soldiers and had finally succeeded in instilling in them a genuine fighting spirit. In August he went onto the offensive, attacking the Austrian positions along the right bank of the Meuse. Bayard's Rock at Anseremme was nicknamed 'slaughter mountain' by the Austrians, so heavy were the losses they sustained there. And at Coutisse, south-west of Huy, the valiant patriots allowed themselves the luxury of sacking the Imperial camp.

BELGIUM, PATRIOT ARMY (I)
1. Mounted Brussels volunteer. — 2. Officer of the Flanders Dragoon Regiment. — 3. Soldier of the Brabant Volunteers. — 4. Brussels Regiment. — 5. Antwerp Regiment. — 6. Drum-Major of the Mons Volunteers. — 7. Officer of the Hainault Dragoons. — 9. Dragoon of the Tongerloo Regiment. — 10. Dragoon of the Namur Regiment.

---

[1] See colours of the Austrian infantry, p. 103.

This run of success induced Congress to order a general offensive, and the languid Schönfeld had to abandon high society once more. To reinforce the regular army, Congress decreed on 28 August the general mobilization of all volunteers, 'military societies'[1] as well as peasants released by the ending of the harvest, for a period of three weeks. The country people involved in this 'peasants' crusade' were commanded by their local clergy and lords of the manor, reinforced by a number of violent monks who preached a 'holy war' with fanatical fervour. Some 15 to 20,000 of these unfortunate people were duly rounded up, divided into companies, 'trained', and on 22 September thrown into battle.[2]

At first there were some encouraging successes: one of Koehler's columns overran several strongpoints with the bayonet. But an ammunition wagon exploded just as the patriots were on the point of breaking the enemy line. The great sheet of flame set fire to the ammunition pouches of some of the nearest attackers, who at once broke and fled in terror. Soon the panic spread to the entire column, and an Austrian counter-attack set the seal on the catastrophe.

During this time, the Prussian Schönfeld had made no worth-while contribution. In fact he was an agent of Frederick II, concerned only with carrying out the orders he received from his master, who had already settled the future of Belgium through diplomatic channels and was now supporting the Austrian restoration. Schönfeld retired in disorder, abandoning all his artillery and supplies, and leaving the road to Brussels wide open. Koehler, the officer and gentleman, fought his way back to Mons, then in obedience to the orders of Congress prepared to defend the capital with his 5,000-odd survivors. But the members of Congress took fright and fled and all resistance melted away.

Joseph II died the same year: 'Your country's finished me,' he told the Prince of Ligne. His successor, Leopold II, offered the Belgians a total amnesty and returned their traditional rights.

---

[1] Societies of Crossbowmen, Archers, Arquebusiers, etc., whose members were peaceful middle-class citizens with more talent for banquets and parades than for fighting.

[2] Known by the Austrians as the Battle of Falmagne, from a village lying between Dinant and Givet.

A year later, the triumph of the French Revolutionary armies brought the annexation of the Belgian provinces to the Republic. Belgium had merely changed masters for a brief space: in the summer of 1793 the Austrians marched back in triumph. After the French regained control of the country in 1797 the introduction of conscription sparked off a general rising, known as the 'Peasants' War' or, in Luxembourg, the *Kloppelkrieg* ('war of the sticks'). It was savagely suppressed.

The politics of appeasement practised by the First Consul of France were unsuccessful in stifling Belgian nationalism, but the quest for liberty – or simply the vagaries of history – led the Belgian people along divergent courses. They gave 12,000 soldiers to Napoleon's Empire, more than thirty of whom became generals, but an even greater number of Belgian generals fought under the Austrian flag.

BELGIUM, PATRIOT ARMY (II)
1. Battling monk of the 'peasants' crusade' forces, raised for three weeks to support the patriot army.—2. England Regiment, also known as the Belgian Legion.—3. Soldier of the Liège Volunteers.—4. Partisan of Dumonceau's Company. These volunteers were men who had been rejected on grounds of inadequate physical stature or fitness when the 1st (Namur) Regiment was raised, and later banded together under the orders of a bold partisan, Jean-Baptiste Dumonceau. Dressed in coats of unwanted yellow cloth, these 'canaries' showed plenty of fight, especially when storming the ruined castle of Poilvache on the Meuse. Their battle-cry was 'Djo, djo!', approximately the Walloon equivalent of 'Let's go!' Their leader was later to make a name for himself as a general in Napoleon's army.—5. Foreign legion known as the 'Turnhout Legion'.—6. Officer of the Tournai Regiment.—7. Brussels Volunteer of the Society of St Christopher.—8. Flanders (8th) Regiment.

Italy

Eighteenth-century Italy was ruthlessly sacrificed to the European balance of power. The rule of Spain was now replaced for the most part by that of Austria, since the fall of the Roman Empire Italy had lost every trace of political unity or independence, and was no longer a nation at all in any real sense.

The War of Spanish Succession had delivered the areas around Milan and Mantua into Austrian hands, as well as Monferrato, which Austria then ceded to the House of Savoy. The treaties of Utrecht (1713) and Rastatt (1714) brought Sardinia and the Kingdom of Naples as well into the Austrian fold. A few years later the Emperor of Austria swapped Sardinia for Sicily, the latter having been the property of the Duke of Savoy, who now styled himself King of Sardinia.

Overcoming his natural apathy, Philip v of Spain mounted an expedition to try to recover Sardinia and Sicily in 1717, but without success. In 1734 his son Charles succeeded where Philip had failed and became King of the Two Sicilies,[1] though in the process he lost Parma and Piacenza to the Austrians.

The War of Austrian Succession (1740–48) gave the Bourbons in France, Spain and Naples the long-awaited chance to peel themselves a few leaves off the 'Italian artichoke' at the expense of Maria Theresa of Austria. The Empress lost only Genoa, but the peace of Aix-la-Chapelle confirmed the Infante of Spain, Philip, as the hereditary ruler of Parma and Piacenza. And so, in the course of the eighteenth century, Italy became the property of the houses of Habsburg-Lorraine (Austria), Bourbon and Savoy, except only for the Vatican States, Modena, and the Republics.

The French Revolution tore Savoy and Nice from the grasp of the King of Sardinia, but was prevented from going any further by the resolute Piedmontese.

The army of Charles Emmanuel III, King of Savoy from 1770 to 1773, was the original model for the later Italian army. He had at his disposal five regiments *di ordonanza* – the Guards, Savoy, Monferrato, Piedmont and Saluzzo – and a regiment of *fusilieri*, named in 1776 the Aosta Regiment. There followed the Chablais Regiment, a battalion of marine infantry, the Queen's and Sardinia battalions and twelve *provinciali* regiments: those of Genoa, Moriana, Ivrea, Turin, Nice, Mondovi, Vercelli, Asti, Pinerolo, Casale, Novara and Tortona. Charles Emmanuel also followed the example of the major powers in having various foreign regiments: the Royal German, and three Swiss regiments from Valais, Berne and Grisons.[2]

His cavalry consisted of seven regiments: King's Dragoons, King's Light Horse, Piedmont Dragoons, Royal Piedmontese, Savoy Cavalry,[3] Sardinian Dragoons and Queen's Dragoons. The light and field artillery were incorporated in the

ITALY
1. Sergeant of the Sicilian Regiment (Piedmont, 1744).—2. Drum-Major of the 'Legione Truppe leggiere' in 1787. This was the first unit of the future 'Guardia di Finanzieri', attached to the frontier guard as an anti-smuggling force. Ten years earlier their colours (visible on the coat facings) had been pale blue.—3. Swiss guard in the Piedmontese service, 1750.—4. Engineer corps (Piedmont, 1745).—5. Ensign of the Naval Regiment (Piedmont, 1744); for reasons of space the size of the standard has been reduced by half.—6. Chasseur-guardsman (Piedmont, 1744).—7. Grenadier of the Monteferrato Regiment (Piedmont, 1744).—8. Savoy Regiment (Piedmont, 1744).—9. Corporal of the Naval Regiment (Piedmont, 1744).—10. Orderly ensign of the Piedmont Dragoons, 1775.—11. Royal Regiment (Piedmont, 1743).—12. King's Dragoon Regiment (Piedmont, 1743).—13. Flag of the Tuscan Regiment in the Austrian service during the Seven Years' War. The flag, of Austrian pattern, has the arms of the House of Tuscany in the centre. -- 14. Artilleryman (Piedmont, 1750).

---

[1] His bizarrely named kingdom consisted of Sicily itself and the Kingdom of Naples. It was Alfonso v of Aragon, King of Sicily, who conquered the Neapolitan peninsula in 1442 to become the first King of the Two Sicilies.

[2] The Swiss served the Bourbons in Naples from 1731, the Venetian republic until 1719, Sardinia from 1243, the House of Savoy (with scores of regiments), the Genoese republic from 1573 until 1779, and the Holy See from 1478 until the present day. The Swiss Guard at the Vatican is recruited by private contract with the permission of the Swiss Confederation.

[3] The famous 'Savoia Cavalleria'. See *Arms and Uniforms of the Second World War*, Vol. III, p. 25.

infantry, but there was also an independent artillery corps with a strength of 2,400 men.

At the end of the century the Kingdom of the Two Sicilies boasted an army of 25,000 ill-trained and ill-disciplined foreign mercenaries. The quality of the puny militias of Genoa and Venice was even more indifferent, while the duchies of Milan and Mantua had had no soldiers under arms for many years.

Spain

One month before his death, in November 1700, Charles II – last of the Spanish Habsburgs – named as his successor Louis XIV's grandson, Philip, Duke of Anjou. The Emperor Francis I of Austria supported the candidature of his own son, the Archduke Charles, and Philip of Anjou's accession thus unleashed the so-called War of Spanish Succession, which was to last from 1701 until 1713.

Hostilities began between, on one side, France, Spain, Bavaria and Portugal, and on the other Austria, supported by Hanover and Brandenburg. But the conflict spread rapidly, to attain European proportions with the entry of England and Holland in support of the Austrians. These powerful reinforcements for the enemy forced France onto the defensive and in 1704 she stood on the brink of disaster.[1] There were setbacks in Italy, and in the naval battle at Vigo (1702); then in Spain, with Portugal's defection from the French side (1703); Gibraltar fell in 1704, and, worst of all, Philip V was driven from Madrid in 1706 by the Archduke Charles, who thus began his admittedly brief reign as Charles III of Spain.

This long chain of reverses was hardly balanced by the occasional French successes at Luzzara and Friedlingen in 1702, Höchstädt in 1703 and Almansa (1707),[2] for crushing defeats at Blenheim (1704), Ramillies (1706), Oudenaarde (1708) and Malplaquet (1709) brought France to her knees and forced the Sun King to sue for peace. The humiliating terms offered by the

SPAIN, BOURBON INFANTRY
1. Ensign, 1700–18.—2. Fusilier, 1700–18.—3. Grenadier, 1700–18.—4. Grenadier, Brussels Regiment, 1718–50 (see also fig. 35).—5. Ensign of the Zamora Regiment, 1718–50.—6. Drummer of the Granada Regiment, 1718–50.—7. Fusilier of the Baza Light Infantry Regiment, 1718–50.—8. Fusilier of the Catalonian Light Infantry Regiment, 1718–50. The breeches are here gathered on the thighs, probably to allow greater ease of movement; they could be let down for protection against cold. The restricting coat is here thrown over the shoulder.—9. Sergeant of the Galician Regiment, 1750–9.—10. Grenadier of the Ultoria Regiment (see also fig. 35).—11. Fusilier of the Catalonian Volunteer Regiment, 1761.—12. Grenadier of the Trujillo Regiment of provincial militia, 1761 (see also fig. 35).—13. Grenadier of the Zamora Regiment (see also fig. 35).—14. Fusilier of the Cordoba Regiment, 1766.—15. Fusilier of the 2nd Catalonian Volunteer Regiment, 1766.—16. Ensign of the Milan Regiment, 1768.—17. Drummer of the Irish Regiment, 1768.—18. Urban militia, 1775.—19. Grenadier of the Zamora Regiment, 1775.—20. Fusilier of the Guards Infantry, 1775–8.—21. Fusilier of the 1st Catalonian Volunteer Regiment, 1775–8.—22. Fusilier of the Asturis Regiment, 1780–9.—23. Grenadier of the American Regiment, 1780–9 (see also fig. 35).—24. Fusilier of the Vitoria Regiment, 1780–9.—25. Grenadier of the Spanish Guards, 1789 (see also fig. 35).—26. Line fusilier, 1789.—27. King's Grenadier, 1789 (see also fig. 35).—28. Grenadier of the Swiss Infantry, 1789.—29. Grenadier of the Walloon Infantry, 1786 (see also fig. 35). An Italian infantry regiment presented the same appearance but with red colours instead of blue. A crowned escutcheon similar to that visible on the cartridge pouch in fig. 28 appeared on the *flamme* of the hat (in the upper centre), as with figs. 28 and 29.—30. Fusilier, 1700–18.—31. Grenadier, 1700–18.—32. Drummer, 1700–15.—33. Fusilier, 1718–50.—34. Grenadier, 1718–50 (see also fig. 35).—35. Grenadier, 1761. This illustration shows the exact arrangement of the *flamme* on the Spanish grenadiers' hats.—36. Fusilier, 1761.—37. Grenadier, 1789.—Figs. 30 to 37 depict the Walloon Guards.—All these drawings are based on Lt-Gen. Condé de Clonard's *Album de la Infanteria española* (Madrid, 1861).

---

[1] This was the heyday of Marlborough (see Vol. I, pp. 92–6).
[2] Off the Spanish Atlantic port of Vigo Spanish galleons under the protection of a French squadron were sunk by the Anglo-Dutch fleet. . . . At Luzzara (in Emilia, Italy) a Franco-Spanish force defeated Prince Eugène's Austrians. At Friedlingen (in western Germany, on the Rhine north of Basle) Villars defeated the Imperial troops. . . . There were two battles at Höchstädt: that of 1703 ended in the defeat of the Austrians by Villars; the second, a year later, is the battle known as Blenheim (see Vol. I, p. 94). . . . At Almansa in Murcia, Spain, the English were beaten by that strange character the Duke of Berwick, son of James II of England and Arabella Churchill (Marlborough's sister) – and a Marshal of France! . . . The battles of Oudenaarde, Ramillies and Malplaquet have already been frequently mentioned.

enemy drove Louis XIV to continue the desperate struggle.

Now fate and the complex game of politics played their parts: the first causing the death of the Emperor Joseph I in 1711, the second eclipsing the brilliant star of Marlborough. The situation was reversed: England pulled back, while the victory of Villaviciosa[1] gave Philip V a chance 'to sleep that night on a bed of enemy flags'. Joseph I's successor, his brother the Archduke Charles (Charles VI), continued the fight in France, but was forced to beat a retreat at Denain in 1712.[2] This latest French success ended the war.

It broke out again in 1717 with the Sardinian and Sicilian expeditions, and more seriously in 1739 against England, who had been looking askance at the illegal competition offered her trade by the Spanish merchant navy. Then came the War of Austrian Succession, during which the Spanish troops, fighting as France's allies in Italy, scored some notable successes.

After several years of peace Spain was forced into the Seven Years' War by Britain's declaration of war in 1762. Though the war was almost over when she entered it, Spain's reward was the loss of Havana, Manila and Florida, for the moment at least. But she received a handsome compensation from Louis XV: Louisiana.

Next, Spain was able to take advantage of the American War of Independence (1775–81) to recapture Florida, though she offered no direct aid to the American rebels and did not even recognize their claim to independence.

The French Revolution found Spain under arms yet again, this time at the instigation of Charles IV, a cousin of Louis XVI, who declared war on the French regicides. This conflict petered out in 1795 after a series of largely defensive encounters, and now Spain joined forces with her recent enemy in yet another war with England. The dark days that marked the beginning of the nineteenth century in Spain have been powerfully captured in the paintings of Goya.

Prior to 1714 there had been 131 Spanish infantry regiments, including two of the *Casa real*

(the Royal Household) eighty-seven national and thirty-six foreign, these last including no less than twenty-seven Walloon regiments. After the War of Spanish Succession these regiments were reduced to 108, a reduction effected largely by reorganizing the foreign troops.

The cavalry in 1703 consisted in ten line and three dragoon regiments, to which was added in 1705 the 1st Regiment of *de la Muerte* (death's-head) hussars. By 1763 there were twelve line and eight dragoon regiments and by the end of the century there were sixteen regiments which together with the Guards and the carabineers of the royal brigade (formed in 1730) represented a total of 10,952 mounted men under arms.

The Walloon Guards

This infantry force, founded in 1702 for the personal service of the King, had a brilliant record under Philip V. But it aroused jealousy elsewhere and was reduced in strength when peace came, most of its officers being discharged. This gradual process of erosion continued from 1706 until the end of the century, the original superb fighting force being seriously diluted by the drafting of numerous foreign troops in 1783. By the beginning of the nineteenth century there remained only a skeleton of the original Guards, and this remnant, too, was destroyed in the course of a last attack during the constitutional revolution of 1822.

---

[1] A small town in Spain, in the province of Guadalajara.
[2] A town on the Escaut, near Valenciennes, where Villars beat Prince Eugène's Austro-Dutch force. See p. 72, note 2, for details of Prince Eugène.

SPAIN, BOURBON CAVALRY
1. Bodyguard, 1703.—2. Dragoon, 1709.—3. Musketeer of the Guards, 1703.—4. Dragoon, 1705.—5. Death's-head Hussar, 1705.—6. Horse grenadier, 1737.—7. Standard-bearer of the Numantia Dragoons, 1739.—8. Line cavalry and Queen's Regiment, 1763.—9. Lusitanian Dragoons, 1763.—10. Dragoon of the Sagunto Regiment, 1775.—11. Line cavalry, Alcantara Regiment, 1775.—12. Royal Carabineer, 1775.—13. Ceuta Lancer, 1789.—14. Line cavalry, Farnese Regiment, 1789.—15. Marie-Louise's Carabineers, 1793.—16. Spanish hussar regiment, 1795.—All these drawings are based on Lt-Gen. Condé de Clouard's *Album de la Caballeria española* (Madrid, 1861).

RUSSIA AND SWEDEN

Our panorama of the armies that fought in the 'lace wars' has now reached the very frontiers of Europe, where two more nations add their voices to the (often strident) harmony of modern Europe: the Russia of Peter the Great and Catherine II, and the Sweden of Charles XII, a country believing in its great-power status and prepared to match itself, recklessly, with its colossal neighbour.

Before 1700, Russia's troops totalled some 130,000 men, including 30,000 cavalry, 63,000 infantry (making up the so-called 'soldiers' regiments') and 20,000 *streltsy*.

The Streltsy

The *streltsy* (from *strielets*, an archer or marksman) were a national guard, comparable to the Turkish janissaries, and founded in 1550 by Tsar Ivan the Terrible.[1] For a long period they represented Russia's only armed force. Their pay was supplemented by a number of special privileges, and as a result they gradually came to be a greater threat to their own master than to the enemy.

After the massacre of the supporters of the young Tsar Peter in 1682 in favour of his half-sister Sofia, who was thus able to establish herself as regent – the *streltsy*, insolent and drunk with pride, set themselves against both church and

---

[1] Ivan IV, 'the Terrible', regarded this nucleus of a standing army as a way of avoiding the need for armies in the feudal mould. The creation of the Streltsy comes under the heading of the centralizing reforms carried out by the first Russian Grand Duke to assume the title of Tsar. He also created a special political police force, the Oprichnina, consisting of a thousand horsemen dressed from head to foot in black and mounted on black horses. Their device of a dog's head and broom symbolized the motto 'To sniff out conspiracy and sweep away treason'. This was certainly the first authentic Russian uniform.

RUSSIA, SENIOR OFFICERS AND GENERALS
1. Senior infantry officer in 1720. — 2. Senior infantry officer in 1742. — 3. Senior infantry officer in 1756. — 4. Senior infantry officer in 1763. — 5. General officer in 1780. — 6. Senior infantry officer in 1786. — 7. Staff officer in 1786.

state. Sofia ordered the execution of their leader, Khovanski, and thirty-four of his most offensive henchmen and twelve of the Moscow regiments were exiled to the frontier territories.

In 1689, correctly judging that her position was threatened, the regent appealed to her 'faithful *streltsy*' to put on a repeat performance of the 1682 massacre, this time including the seventeen-year-old Tsar. The *streltsy* refused point blank, one regiment going so far as to put itself under the orders of the young Tsar. Aware that her reign was over, Sofia retired gracefully to a convent, and Peter I – Peter the Great – made a triumphant entry into Moscow between two columns of 18,000 *streltsy*.

During one of his celebrated journeys round Europe, in 1698, the Tsar learned that a revolt by the *streltsy* had been put down in his absence, those versatile opportunists having spread the rumour of Peter's death and tried to restore the former regent to the throne. Haunted by the memory of the massacre of 1682, Peter decided to exterminate this 'bad seed of the Russian soil'. Within a few weeks 200 *streltsy* were decapitated at Preobrazhenskoy[1] and more than 800 at Moscow. It is even said that the fall of the 1,000th head was celebrated by a lavish banquet. It is a fact that the Tsar accounted for five with his own hands, and that his officers and state dignitaries, 'invited' to the gruesome event, bagged 109 on 17 September. The survivors, deprived of their arms and exiled into the provinces, had to readapt themselves to civilian life.

The uniform worn by the *streltsy* from the beginning had consisted of a long caftan, its colour varying according to regiment, with horizontal stripes of black or pink braid. Headgear was a conical cap of varying colour, trimmed with expensive fur. Their principal armaments were the arquebus and the *berdych*, a kind of short halberd with a blade like an elongated axe on which the arquebus was rested when shooting. The *berdych* became obsolete with the introduction of the lighter musket.

[1] Village near Moscow, the site of their 'exploit' in 1682.

RUSSIA, GUARDS (I)

1. Grenadier of the Semionovski Regiment in 1700. — 2. Musketeer of the Preobrazhenski Regiment in 1700. The grenade pouch of No. 1 had the double-headed eagle in the centre and four grenades in the corners, while No. 2's had only the oval monogram which can be seen on the cartridge pouch in fig. 1. — 3. Musketeer of the Preobrazhenski Regiment in 1761. The Semionovski Regiment was identical at this period except for its light blue collar. A third regiment, the Ismailovski, had green coats and collars. — 4. Musketeer of the Semionovski Regiment in 1720. The red tunic might sometimes be replaced with blue when imported cloth was in short supply. In 1742 the tunic was shortened, while the hat became wider and acquired tassels at the sides. — Musketeer officer of the Preobrazhenski Regiment in 1730. — 6. Grenadier of the Ismailovski Regiment in 1742. In ordinary dress the gaiters were black, and were worn over white knee covers. N.C.O.s wore the same uniform with gold braid at the neck and round the cuffs. — 7. Grenadier of the Preobrazhesnki *Leibkompanie* in 1761. This was the regiment's former grenadier company, which was honoured with the title of *Leibkompanie* in recognition of the part it had played in the events that took place when the empress first came to the throne. — 8. Flag of the *Leibgarde* in 1762. — 9. Chevalier guard in 1724. The satin-stitch of the hat was white mingled with red; the surcoat worn over the uniform coat had a silver St Andrew's star in the centre, and the star in turn had at the centre a blue St Andrew's cross on an orange background surrounded by a circular sky-blue banderol. The back displayed the black double-headed eagle, with golden talons, beaks, sceptre and terrestrial globe.

Infantry of the Guards

In 1700 this was still the only unit with a standardized form of dress. The nucleus of the two regiments of the Guards were the children's regiments, the *potyechnii*, with whom Peter had played at war games since his childhood. Having seen off the intrigues of the regent Sofia, Peter, then aged seventeen, entrusted his 'regimental playmates'[1] to his faithful Scottish friend Patrick Gordon with instructions to train them in the European fashion.

The two elite regiments thus established were called respectively Preobrazhenski and Semionovski, after the villages in which they originated, and were placed under the command of two Princes, Romonadovski and Buturlin,[2] who happened to detest one another. Their first 'grand European-style manoeuvres' degenerated into a pitched battle, with scores of dead and hundreds wounded. The post-mortem developed into a brawl between the two noble commanders; the unfortunate Gordon, attempting to intervene, was promptly beaten up. Calm was eventually restored by generous application of the *dubina*, Peter's ivory-handled stick.

A third regiment, the Ismailovski, was raised by the Empress Catherine I. She recruited Ukrainians, Estonians and Courlanders; of necessity she also included some Russians, though these were regarded as unreliable. The command was entrusted to a Baltic noble, von Löwenwald. A number of Germans, including Colonel Christoph Hermann von Manstein,[3] also served in the Guards.

Catherine I, Anna Ivanovna the 'Colonel of the Preobrazhenski', Elisabeth Petrovna and Cather-

---

[1] Also known early on as his 'toy battalions'.
[2] The Tsar's companions in debauchery, and a pair of out and out rogues. The first was a former supremo of the *potyechnii* and head of the secret police; the second, known as the 'Tsar of Semionovskoy', was a former colonel of the Streltsy.
[3] An ancestor of Erich von Lewinski von Manstein, Marshal of the Third Reich.

RUSSIA, GUARDS (II)

1. Chevalier guard in everyday dress with surcoat, in 1764. — 2. Chevalier guard in full dress, 1764. The golden two-headed eagle at the centre of the star on the surcoat was studded with diamonds and sapphires. The same motif, without the jewels, was repeated on the back. — 3. Chevalier guard in campaign and full-dress uniform, 1742. From 1730 to 1741 the tunic was chamois, and had the same braid as the coat. The rapier was carried instead of the straight sabre. From 1742 the hat became wider and the gold oval monogram in the centre of the cuirass displayed the initials of Elisabeth Petrovna instead of Anna Ivanovna. — 4. Horse guard in dragoon-style ordinary dress, 1760. — 5. Trumpeter of the Horse guards in 1763. — 5a. False sleeve (detail). — 6. Musketeer of the Preobrazhenski Regiment in 1742, on sentry duty. — 7. Grenadier of the Ismailovski Regiment in 1730. — 8. Musketeer of the Semionovski Regiment in 1742. Apart from the regimental colours visible at the neck, the uniforms of the other regiments were identical. — 9. Horse guard in ordinary dress, 1786. The plume was not introduced until 1781; the campaign and full-dress uniform had remained the same since 1742 (see fig. 3). — 10. Musician of the Guards infantry from 1742 to 1761. He belonged to the Semionovski Regiment, as can be seen from his light-blue collar. — 11. Drummer of the *Leibkompanie*, 1742–61. The central feathers were white. — 12. Helmet of the mounted guards. N.C.O.s had black and white plumes; officers all white — 13. Mitre of a grenadier of the *Leibkompanie*. The plume was worn only for parades and guard duty: they were all red for private soldiers, white on the outside and red at the centre for N.C.O.s, and all white for officers.

ine the Great could always count on the unswerving loyalty of these regiments. Elisabeth Petrovna commemorated her difficult accession to the throne by founding the *Leibkompanie* in 1742 from those former grenadiers of the Preobrazhenski Regiment at whose head she had marched on the Winter Palace. On the other hand, the Tsars Peter III (1762) and Paul I (1796–1801) were assassinated by officers of their own Guards. The Preobrazhenski Regiment, the most illustrious of all, was destined to play a leading part in most of the tragedies that disfigured the last two centuries of Imperial Russia.

In the reign of Paul I, the demented son of Catherine II, the Guards were inspected meticulously by their sovereign every day, even if there were a hundred degrees of frost!

Cavalry of the Guards and Horse Guards

The first cavalry unit in the Russian Guards was no more than a mounted escort given to Catherine I on the occasion of her coronation in 1724.

The chevalier guards of the *kavalergardia*, officers to a man, had the sovereign for a 'captain'. The lowliest guardsman had the rank of lieutenant, and the corporal was a lieutanant colonel! Disbanded, reformed and then disbanded again in the space of seven short years, these chevalier guards (or rather their name) reappeared in 1742, but only in the most exceptional circumstances, when their uniforms were worn by the grenadiers of the *Preobrazhenski Leibgarde*.

Catherine the Great reintroduced an authentic cavalry unit into the Guards in 1762. Her son Paul I suppressed it, then reintroduced it as a regiment in 1799.

Another mounted force, founded as the *Leibregiment* in 1707, was known under an assortment of

[1] The German word *Leib*, literally 'body', was used first in Germany, then in Russia and elsewhere, to signify an elite corps commanded by the monarch in person; *Leibgarde* should not, therefore, be translated 'body guard' or 'life guards', which means simply a personal escort provided for a king or prince.

RUSSIA, INFANTRY (I)
1. Officer in 1700.—2. Musketeer N.C.O. in 1700.—The colour of the coat was no indication of rank, but was worn throughout a regiment. Early in the eighteenth century the famous dark green had not yet become dominant, and a wide range of colours was on display, depending on the whim of the commanding officers, or simply on what happened to be available. Different coloured coats are sometimes met with within a single regiment. The *pokalem* hat, with its flaps which could be pulled down in winter, offered a wide range of contrasting colours, and was worn instead of a hat by many regiments until 1720.—3. Musketeer in 1700.— 4. Grenadier in 1700.—5. Musketeer in 1700, wearing the *pokalem*.—6. Musketeer, 1730.—7. Grenadier, 1730.—8. Junior officer of grenadiers, 1730.—9. Musketeer, 1720.—10. Musketeer, garrison infantry, 1720.

different names before taking the title of 'Horse Guards Regiment' in 1731.

Paul I, whose insanity did not stop him ruling Russia for five years, took it into his head on one occasion to drill the famous regiment himself. Mumbling an indistinct command, he set off at a gallop, then suddenly wheeled and retraced his steps, with an enraged scream of 'Regiment, by squadrons, to Siberia, at the walk . . . *march!*' It took hours of pleading to pacify the Tsar and re-assemble the regiment, which had understood nothing of what its ferocious master had said but the word 'Charge!' Terrified and bemused, the unfortunate horse guards were already sixty miles from Moscow!

The Line Infantry

The uniform of a common infantry soldier under Peter the Great presented a kaleidoscope of almost all the classic colours fashionable in Europe's armies. Russian green was already in evidence, but frequently mingled with white, red, yellow and blue, a phenomenon caused by difficulties of supply.[1] The bizarre *pokalem*, which enjoyed an extraordinary popularity as far away as Napoleonic France, marched side by side with the mundane western hat. The distinctive pocket-flaps, with their five 'crenellations', were the only authentically Russian element of the uniform.

In about 1708 there were some fifty regiments of line infantry, including five of grenadiers, most of them known by the name of a province rather than that of their colonel: these names died out around 1730. There were forty-five line regiments in 1763; forty-seven in 1765; sixty-three in 1777; fifty-nine in 1786; and fifty-seven in 1795. The grenadier regiments, however, increased from four (1763) to ten (1786) and then fifteen (1795).

The multicoloured uniforms of the old days ceased to exist in 1730, as did the cockade – actu-

[1] There were only sixty-four textile mills in Russia by the end of Peter the Great's reign.

RUSSIA, INFANTRY (II)
1. Musketeer in 1760. The red gaiters mark him as a member of the Apcheronski Regiment, which earned this unique distinction by fighting 'up to its knees in blood' at the Battle of Kunersdorf in 1759. The stance is that following the order, 'Ground arms!'.—2. Grenadier with arms reversed (for funeral duty), 1756. His mitre-helmet, with its impressive neck-guard, is the model introduced in 1756, and was abandoned in 1759 in favour of the old cloth mitre. The cartridge-pouch was covered with a copper plate bearing the regimental arms.—3. Musketeer sergeant in 1756 in praying attitude.—4. Junior officer of musketeers in 1756. He is covering the firing-mechanism of his gun, as was the rule in wet weather. This weapon seems to have replaced the traditional spontoon in 1734; the ammunition needed for the new weapon was kept in a small pouch on the belt similar to those in figs. 2 and 8.—5. Musketeer of the Guard (Ismailovski Regiment, as can be seen from the colours at the neck) executing the first movement of the marching salute, reserved for officers.—6. Grenadier of the Guard (Preobrazhenski Regiment, see red collar) in 1762, executing the second movement of the same salute.—7. Junior officer of the grenadiers of the Guard (light-blue collar, therefore Semionovski Regiment) performing the third movement of the marching salute.—Junior officer of the line grenadiers in 1756, at the 'shoulder arms'.—8a. Profile view of the 1756 model mitre-helmet.—9. Junior officer of the line musketeers in 1762, at the salute. The green of his coat is a darker colour than hitherto, and easily distinguished from the light green of the Guards (see figs. 5, 6 and 7). Tunics and breeches are in varying shades of yellow. The metal of braid and buttons varied according to the regiment.

ally a white bow – previously worn by the Guards.

Peter III reigned for only a few months, but still found time to disguise his infantry as Prussian soldiers,[1] in the tradition of the small army he had organized in his capacity as Duke of Holstein-Gottorp, before his accession. His wife, Catherine II, presided over the extraordinary changes made to the uniforms by her favourite, Potemkin; but his son, Paul I, a fervent admirer of Frederick II, made it his first priority to reintroduce the Prussian-style uniforms.

The Potemkin Uniform

It was the imagination of Prince Potemkin[2] that, in 1786, led to the introduction of a completely new uniform – practical and, in some respects, a century ahead of its time. This uniform had the additional merit of being virtually 'all purpose': it could be worn by all units except the Guards, the Cossacks and a few special troops.

The heavy and cumbersome hat was replaced by a helmet with peak, horsehair crest and cloth flaps which would be tied down on the neck to protect the ears from cold. The short tunic, known as the *kurtka*, was well cut, and the trousers had fixed gaiters of ample width, much more pleasant to wear than skimpy breeches and tight-fitting gaiters. Also the hair, previously tied at the back, was now cut short and left unpowdered by these line troops.

---

[1] At least in part, as with his artillerymen.
[2] Grigori Alexandrovich Potemkin (1739–91) had certainly played a part in Catherine II's coup d'état in 1762. After a dazzling career during the first Russo-Turkish War (1768–74) he became the Empress's favourite and is associated both with her great internal reforms and her foreign policy of southward expansion, involving such achievements as the exploitation of the Ukrainian steppes, the annexation of the Crimea, and the establishment of urban centres, an arsenal and naval dockyards. He founded Sevastopol, the base for the new Black Sea fleet. The triumphal progress through the Ukraine which he organized for Catherine II in 1787 marked the apex of his career. But as Commander in Chief of the Russian forces at the outbreak of the Second Russo-Turkish War (1787–91) he lost his fleet in a storm – and lost with it the favour of the Empress.

RUSSIA, INFANTRY (III)
1. Drummer, 1756. – 2. Sergeant of musketeers, 1762. – 3. Grenadier, 1762. – 4. Drummer, 1763. – 5. Sergeant of musketeers of the Guards infantry (Preobrazhenski Regiment) in 1763. – 6. Musketeer of the Guards infantry (Semionovski Regiment) in 1763. The third Guards regiment, the Ismailovski, differed from the others by having a three-coloured plume, black at the top, red in the centre and white at the base; their hats had green tassels at the sides, and their coat collars were of the customary green colour while their yellow shoulder-flaps had the usual monogram in green. – 7. Drummer of the chasseurs, 1777. – 8. Sergeant of chasseurs, 1763. – 9. Chasseur, 1763. – 10. Chasseur of the Guards (Preobrazhenski) in 1770. – 11. Chasseur of the Guards (Semionovski) in 1770. – 12. Chasseur of the Guards (Ismailovski) in 1770. – 13. Chasseur of the Guards (Preobrazhenski) in 1786. – 14. Chasseur of the Guards (Semionovski) in 1786. – 15. Chasseur of the Guards (Ismailovski) in 1786. – 16. Drummer, 1786. – 16a. Trouser braid (detail). – 16b. Helmet *flammes* seen from behind. – 17. Musketeer, 1786. – 18. Grenadier, 1786. – 19. Chasseur, 1786. – 20. Chasseur, 1797. – 21. Grenadier officer, 1797.

Line Cavalry

The Dragoons

The first regular mounted troops appeared on the scene during the reign of Peter the Great. These dragoons, like the infantry, until 1720 were dressed in a variety of colours, for reasons already noted, but in 1720 this motley plumage gave way to a blue uniform.

The dragoon regiments had numbered about thirty at the beginning of the century. In 1712 four were disbanded, but the total was increased again to thirty-nine in 1741. The introduction of other types of cavalry caused considerable reductions in the strength of the dragoons, who numbered seven regiments in 1763, fourteen in 1765, eight in 1775, ten in 1786 and finally eleven in 1796.

The Dragoon Grenadiers

There existed a few regiments of dragoon grenadiers, whose uniform was distinguished only by their wearing of the mitre. Later, in about 1745, they accounted for two of the twelve companies in each dragoon regiment. The dragoons' and cuirassiers' uniforms were unaffected by Peter III's reforms of 1762; from 1786 until 1796 they wore the imaginative uniform designed by Potemkin.

The Cuirassiers

This type of heavy cavalry made its appearance late in Russia, in 1731. The initial three regiments were increased to six in 1763, then reduced to five from 1775 until the end of the century. The cuirass, consisting only of the blackened steel breastplate, was dropped when the new Potemkin uniforms were introduced in 1786; it then reappeared together with the old uniform, on the accession of Paul I.

RUSSIA, CAVALRY (I)
1. Dragoon, 1700.—2. Dragoon-grenadier, 1702.—3. Dragoon in 1712, wearing the *pokalem*.—4. Dragoon, 1730.—5. Grenadier of the dragoons dressed for infantry duty, 1730. Apart from the mitre, the design on which was different for each regiment, the uniform is identical to that in fig. 4. Note the inevitable match-holder on the crossbelt carrying the grenade pouch; the other bandolier carries the traditional cartridge-pouch. Having no strap and snap-link, the grenadier carried his weapon slung over his shoulder by its sling.—6. Dragoon, 1756. The grenadiers had the same uniform but wore the mitre-helmet used by the infantry. In 1759, they adopted the hat.—7. Dragoon, 1754.

The Carabineers

In 1763 nineteen carabineer regiments were formed, at the expense of the dragoon and cuirassier regiments. Virtually identical to the dragoons, though regarded as heavy cavalry, their numbers were increased by an extra regiment in 1765. The total fell to nine in 1775, was increased again to nineteen in 1786, and finally levelled out at sixteen in 1796.

The Hussars

Apart from a brief appearance under Peter the Great, these light cavalry were not seen in Russia on any serious scale until 1741, when they were introduced in emulation of the Hungarian hussars in the Austrian service. From the beginning, both the *mirliton* and the fur *kolpak* were worn. The number of regiments varied frequently, as may be seen from the illustrations. In 1783 the hussars ceased to exist – apart from the *Leibhusaren* of the Guards – and became ordinary light cavalry. They began to make a rather hesitant comeback in 1788.

The Lancers

The lancers date back to 1765, when there were four regiments of *pikineri*, distinguishable from one another as follows:

	REGIMENTAL COLOUR	SASH
Dnyeprovski	Green	Black
Donetzki	Sky blue	Black
Elisavetgradski	Red	Yellow
Luganski	Yellow	Black

With the adoption in 1776 of a new uniform designed by Prince Potemkin, the bonnet was replaced by the Polish *konfederatka*. The six regiments of this period were distinguishable as follows:

	REGIMENTAL COLOUR	SASH
Dnyeprovski	Green	Black
Ekaterinoslavski	Sky blue	Black
Elisavetgradski	Pink	Black
Luganski	Yellow	Black
Poltavski	Orange yellow	Black
Khersonski	Black	Yellow

In 1783 all the lancers were incorporated into the light cavalry.

RUSSIA, CAVALRY (II)

1. Dragoon in 1775. — 2. Carabineer in 1763. — 3. Carabineer in 1778. — 4. *Pikineri*, lancer of the Donetz Regiment in 1764. The lance, like the shabraque, was in the regimental colour. The diagram on the left shows the pattern and arrangement of the buttonholes. — 5. Lancer of the Elisavetgradski Regiment in 1776. His hat is the *konfederatka*, of Polish origin. He wears two caftans, the sleeves of the lower one passing through the openings in the sleeves of the outer, which are left to hang free, or knotted behind the back. — 6. Dragoon in 1786. — 7. Carabineer in 1786. The shabraque was red with white edges and decoration. — 8. Light horse (formerly lancers) in 1786. — 9. Mounted chasseur in 1778. The shabraque was dark green, edged with black, with the monogram and laurels in white. — 10. Trumpeter of the lancers in 1776 (Dnieprovski Regiment). — 11. Trumpeter of the carabineers in 1786. — The double *flammes* of the 'Potemkin uniform', shown in figs. 8, 9 and 11 in front of the wearer's shoulders, were in fact of course worn hanging down at the back.

142

Light Cavalry and Mounted Chasseurs

With the new Potemkin uniform there arrived the sixteen light cavalry regiments, whose uniform closely resembled that of the carabineers. Disbanded on the accession of Paul I, they were used in part as a nucleus for the new hussar regiments.

In 1788 sixteen troops of mounted chasseurs were attached to the sixteen light cavalry regiments, whose uniform they adopted – though always in green. A year later they were expanded to four regiments, wearing black *mirlitons* with black *flammes* trimmed with dark green. A single regiment survived into the reign of Paul I, though now converted to hussars.

The Cossacks[1]

The Cossacks, of Slavic origin, were a people living scattered through the European and Asiatic provinces of Russia, north of the Black Sea and the Caspian, in the Caucasus and Siberia.

The best known of them were the Ukrainian Cossacks, and those living in the lower courses of the Don and Volga. In the Ukraine, the turbulent *za porogani* Cossacks – so called because they were settled beyond the falls of the Dnieper – were particularly liable to resent any outside interference with their independence. They lived in a democratic republic, the *sietche*, and cherished a respect for the code of honour that bordered on the ludicrous. Brought under Russian domination in 1654, they revolted successively against Peter the Great, Catherine II and Nicholas I.

RUSSIA, CUIRASSIERS
1. Cuirassier in 1730. – 2. Cuirassier in 1756. – 3. Cuirassier in 1763. – 4. Cuirassier of the Ekaterinoslavski Regiment in Potemkin uniform, 1786. – 4a. Helmet (detail). – 5. Trumpeter, 1745. – 6. Trumpeter, 1763. – 6a. Detail of one of the two embroidered strips at the back of the uniform – they were in fact atrophied loose sleeves, and had the same fawn background colour as the coat.

[1] See *Arms and Uniforms of the First World War*, Vol. II, pp. 46–8.

As may be imagined, it was easier to devise a uniform for these dynamic recruits than to induce them to wear it. Not until 1774 is it possible to discern any clear distinctions of colour. Here are some examples:

	HAT	CAFTAN	COLLAR, CUFFS AND LINING
Don Cossacks	Sky blue	Sky blue	Red
Volga Cossacks	Red	Red	Red
Ural Cossacks	Pink	Light blue	Light blue
Astrakhan Cossacks	Dark brown	Dark brown	Dark brown
Azov Cossacks	Sky blue	Sky blue	Sky blue

	HALF-CAFTAN AND TROUSERS	SASH
Don Cossacks	Sky blue	Pink, yellow fringes
Volga Cossacks	Red	Sky blue
Ural Cossacks	Light blue	Pink
Astrakhan Cossacks	Dark brown	Sky blue
Azov Cossacks	Sky blue	Pink, yellow fringes

In 1788 the first true uniform, including the *kurtka*, gave a foretaste of the style of the Cossack regulars of the Napoleonic period. In western Europe during the eighteenth century – and indeed during the nineteenth – the word 'cossack' came to mean the perpetrator of any short, violent raid ending in pillage, a synonym for brutality and violence. A fair summary of the military virtues of the Cossack.

The Artillery

The early Russian artillery consisted of a host of weapons which had nothing in common except their very poor quality. Into this chaos Peter the Great first attempted to instil some sort of order. Cast iron quickly replaced bronze. Peter's dragoons were allocated the world's first horse artillery, and the varying callibres of the field pieces were reduced to three in number: 3, 3/8 and 4 inch.

The next reform dates from 1750. There now appear the 'secret weapons', the 152-mm howitzers with bell-shaped muzzles to ensure a better spread of shot; then in 1757 come the five new rapid-firing 'unicorns' with calibres ranging from 95 to 245mm. In 1787 the bigger unicorns and the howitzers were discontinued, and further improvements led to the adoption of 'light' and 'medium' cannon, while three classes of unicorn were still produced.

In 1741 there was only a single regiment of artillery, and three corps of siege artillery. But the regimental artillery attached to each regiment gave it between five and eight supporting cannon according to the strength of the regiment (in 1795).

In 1763 the field artillery consisted of one regiment of bombardiers, two of cannoneers and two of fusiliers. From 1787 onwards the field artillery was formed into provisional brigades and in 1794 the first five companies of horse artillery were formed. The fortress artillery, a self-contained force, was as everywhere equipped with an erratic assortment of obsolete weapons. Gun-carriages and wagons were painted red until 1709, then green until 1763, then red again until 1796, and then green again, this time for good.

A *Leibgarde* artillery battalion was established by Peter III in 1762, but Catherine II disbanded it the following year. She gave independent status to the bombardiers of the Preobrazhenski Regi-

RUSSIA, HUSSARS (I)
A: 1. Hussar of the Georgian Regiment in 1742.—2. Officer of the Moldavski Regiment in everyday dress. In full dress the sword-knot was all gold, and gold frogging and trimmings were more lavishly in evidence. All officers had yellow boots, except those in the yellow hussars and the Kharkovski Regiment who wore the same black boots as their men.—3. Trooper of the *Leibhusaren* in full dress, 1776.
B. Hussar regiments from 1741 to 1761: 1. Serbian.—2. Hungarian.—3. Georgian.—4. Moldavian.—5. Lieutenant-General Horvath's.—6. New Serbian.—7. Slobodski.—8. Yellow.—The uniforms of the Gulgarian and Macedonian Regiments and those of Lieutenant-Generals Preradovich and Chevich are unknown.
C. Hussar regiments from 1763 to 1776: 1. Vengerski (Hungarian).—2. Grouzinski.—3. Moldavski.—4. Serbski. —Raised in 1764: 5. Yellow.—6. Black.—7. Bakhmoutski.— 8. Samarski. Raised in 1765:—9. Akhtyrski.—10. Isioumski. —11. Ostrogozhski.—12. Soumski.—13. Kharkovski.—14.— 15. Mirliton with the *flamme* rolled and unrolled.—16. *Leibhusar* of the Guards, 1776.

ment, as well as the artillery sections of the Semionovski and Ismailovski Regiments. In 1796 these three units were used by Paul I to reconstitute the *Leibgarde* battalion.

The numerous different uniforms we have shown – brightly coloured, often original in design – depict the 'classical' units which fought for the eight different Russian regimes of the eighteenth century. As well as these there was a throng of garrison troops and mixed, provisional or independent units: the *Landmiliz*, the provincial companies and sections and the special Gatchina troops – those living toys used by the Tsarevich Paul Petrovich, later the Tsar Paul I, to prepare for the reintroduction of the old Prussian uniforms favoured by his murdered father, as a gesture of hatred for his mother, Catherine the Great.[1]

---

[1] The 'prussomania' evinced by some of the Tsars never ceases to astonish. Peter III, an unintelligent, ill-educated man, was the son of a German prince. His contempt for Russian tradition went hand in hand with an ill-timed admiration for the enlightened despotism of Frederick II and his Prussian-style manoeuvres. As soon as he ascended the throne he recalled his troops, then occupying the majority of Prussia, and went so far as to conclude an alliance with Frederick. Paul I Petrovich inherited his father's ideas, organizing his troops on the Prussian model and marrying two German princesses in succession; his mother, be it remembered, was born Sofia von Anhalt-Zerbst. But Catherine the Great was a very different matter: her political genius allowed her to enter completely into her role as an authentically Russian empress, the heiress of Peter the Great.

RUSSIA, HUSSARS (II)
Regiments from 1776 to 1783: 1. Akhtyrski. — 2. Bulgarian. — 3. Byelorussian. — 4. Hungarian. — 5. Wallachian. — 6. Dalmatian. — 7. Isioumski. — 8. Illyrian. — 9. Macedonian. — 10. Moldavian. — 11. Ostrogozhski. — 12. Serbian. — 13. Slavianski. — 14. Soumski. — 15. Ukrainian. — 16. Kharkovski. — In 1783 and 1784, all the hussar regiments became light horse, leaving only the *Leibhusaren* (see previous page, fig. 16). 1788 (reintroduction of line hussars): 17. Voronezhski. — 18. Olivopolski. — 19. Hussar squadron attached to the Pskovski Dragoons (disbanded in 1796). — 20. Moscow Squadron (attached to the Moscow city police, disbanded in 1800). — 21. Hussar of the Slavianski Regiment in 1765. — 22. Officer of the Olviopolski Regiment in 1788. — 23. Hussar of the Moscow Squadron in 1786.

Sweden

Charles XII

Historians have been loud in their praises of the military virtues of this adventurous, intuitive, impetuous hero-king, but, as Voltaire wrote, 'In Charles XII, the heroic virtues took a form so exaggerated as to become as dangerous as the contrary vices.' Intensely ambitious, and driven by a genuine passion for war, the king whose colossal stubbornness earned him the nickname 'iron head' expected from his troops a performance that often exceeded the limits of human reason. He committed the final reckless error of underestimating his opponent and marching to disaster in the heart of the gigantic Russian empire.

Yet his beginnings had been promising enough. In 1700 Charles attacked the Poles who were laying seige to Riga; Peter the Great responded by besieging Narva.[1] The young King attacked the Russians by having most of his infantry transported to the field on the cavalry horses.[2]

In 1703 Charles took on a combined force of 24,000 Saxons and Poles at Kissow, occupied the whole of Poland, and, in 1706, invaded Saxony. In 1707 he attacked Russia: he crossed the Beretsina, but instead of following up his attack towards Moscow he turned aside into the Ukraine, in response to an appeal from the Cossack hetman Mazeppa. He won a battle, at Golovtsino, but his second in command, Adam Ludvig Lewenhaupt, was defeated at Liessnaya.

---

[1] Town in Esthonia, U.S.S.R. The region had belonged to Sweden since the sixteenth century. Peter the Great was seeking an outlet on the Baltic, before the founding of St Petersburg in 1703; he took the town in 1704, and the whole of Esthonia was absorbed into Russia in 1721.

[2] The sizes of the forces involved vary according to the degree of admiration each source shows for Charles: 4,000, 8,500 or 9,000 Swedes against 60,000, 40,000 or 20,000 Russians at Narva; 16,000, 20,000 or 30,000 Swedes against 60,000 Russians at Poltava.

RUSSIA, COSSACKS

1. Cossack from Little Russia, first third of the eighteenth century. — 2. Zaporogue Cossack. Their heads were shaven, except for a single lock, the end of which was curled around the left ear. Their clothes were sometimes made of blue cloth, or of a mixture of the two colours. — 3. Don Cossack (line) in 1774. The caftan had slashed sleeves to allow the sleeves of the half-caftan worn underneath to pass through, as with the right arm of the man shown here. The sleeves could also be rolled up on the shoulders to show the contrasting lining, rather like the wings in figs. 4 and 5, or alternatively could be knotted at the back. The cossack, the epitome of the light cavalryman, attacked with his lance not tucked under his arm, but held at arm's length where he could feint with it as with a rapier; he could hit his adversary wherever he liked with deadly skill. The sword is the cossack *shasqua* without a guard. — 4. Tchougouyev's Cossacks in 1776, campaign dress. — 5. Tchougouyev's Cossacks in 1776, full dress. The escort troop of the Don Cossacks (Guards), raised in 1775, wore the same uniform but with colours reversed. The shabraque was red edged with white, with the crowned royal monogram and two laurel wreaths picked out in white also. — 6. Ekaterinoslav Cossack, 1789.

In 1709, his fighting strength already considerably reduced, he laid siege to Poltava,[1] where he was attacked in the rear by a Russian army of 50,000 men. Wounded, his army smashed, the young conqueror took refuge in Turkey, leaving behind him 9,000 dead, 3,000 prisoners and Lewenhaupt's decimated force.

The assistance of the French Ambassador enabled Charles to take his revenge in 1711, when he induced the Turks to declare war on Russia. Peter the Great had rashly advanced with his weary troops into Wallachia. Defeated, Peter had to surrender the port of Azov and allow the Swedish troops free passage home.

Charles then attacked Norway; but in 1718 a bullet put an end to his career, and to his country's 'golden age'.

Sweden took part in the Seven Years' War... on the Russian side! She came out of it with her territory intact, having signed a separate peace with Prussia in 1762.[2]

The Uniforms

Since 1690, the Swedish uniform had consisted of a dark blue coat, with regiments being distinguished by the colours of their breeches and stockings.

The general rig-out changed little in the next sixty years, and Sweden entered the Seven Years' War with very outmoded uniforms. After adopting the Prussian style in 1765 the army went in for 'Swedish costume', a comic-opera style that illustrates admirably the almost inevitable failure of any violent changes or attempts at 'futurism' in matters of dress. In 1798, under Gustavus IV, came a return to a more sensible idea, the coat-tunic; but the distinctive tall hat was retained, an unsuccessful effort being made to give it a slightly more military appearance.

---

[1] Town in Ukraine, U.S.S.R. This battle marked the end of Swedish supremacy in the Baltic area, and the entry of Russia into European affairs. Peter the Great was forced to modernize his empire if he wished to win.

[2] The military history of Sweden in the eighteenth century is the best possible illustration of how relative is the idea of the 'hereditary enemy', how vain the ambitions of lesser powers. After Charles XII's day, Sweden fought another war with Russia, in 1741–43 – this one cost her south-eastern Finland. During the Seven Years' War Russia and Sweden found themselves allied again, both having joined the Franco-Austrian cause against England and Prussia. Under Gustavus III (1771–92), an enlightened philosopher-despot, Sweden played its part in the grand cultural scene (Linnaeus, Celsius, Swedenborg), but remained militarist at heart. There was another war with Russia in 1788; then Gustavus IV (1792–1809) fought first against Napoleon, then against Russia . . . and finished up losing Finland. Charles XIII (1809–18) made peace, but in 1810 he named as his successor the French Marshal Bernadotte, who promptly plunged Sweden into the final struggle against Napoleon.

RUSSIA, ARTILLERY AND ENGINEERS
1. Bombardier with grenade-thrower in 1700.—2. Artillery officer in 1729.—3. Artillery fusilier in 1720. His is simply civilian dress, 'militarized' by the use of an easily visible colour.—4. Officer of bombardiers in 1757. The fusilier officer wore the same uniform but with edging on the hat in gold.—5. The corps of engineers wore the artillery uniform but with the buttons in white metal.—6. Fusilier, 1759. The officer's uniform was the same with the addition of the sash shown in fig. 4, braid on the hat as well as the collar, facings and cuffs of the coat, and the pockets and facings of the tunic. —7. Cannoneer in 1762. The officer had gold braid on facings and cuffs.—8. Cannoneer in 1763. White breeches were worn in summer. The fusilier's uniform was similar, but with no lapels and only a single row of buttons.—9. Fusilier from 1786 to 1796. The sapper had the same uniform but with helmet plate and buttons in white metal and a white stripe on the trousers. He was also armed with a sabre.—10. Bombardier battalion of the *Leibgarde* in 1762, disbanded in 1763.—11. Bombardier of the company attached to the Preobrazhenski Regiment in 1763; disbanded the same year.—12. Mounted artilleryman in full dress, with plumed hat, 1786 to 1796. The shabraque was red, edged and decorated with the imperial monogram and the traditional laurels in yellow.—13. Garrison artillery, 1786. Many units of the artillery and engineers wore this cap, which was extraordinary in its day.

Conclusion

The apparel does indeed proclaim the man. Its study may seem dry and unproductive: yet every page evokes memories of glory or disaster, and old cast-offs take on new life to relive a forgotten charge, or crumple in bleeding, anonymous heaps.

The armies of the old days collected all the young hotheads, all the bloody minded, the disreputable, the naive. Discipline became steadily better, with a consequent reduction in the frequency and severity of punishments: France was the first to abolish corporal punishment in the army, on 14 July 1789!

The words 'lace war' conjure up a picture of perfumed marquises, elegant and not always entirely genuine, trying to compensate with tight lips and martial swagger for their beautifully powdered hair and foppish uniforms. But in battle these exquisite, smooth-cheeked young gentlemen showed the most dazzling courage, the first essential of their profession. The extraordinary thing is that the guttersnipe and the rustic could put on blue, red or green and at once become a member of a race apart, ready to dare anything, sacrifice anything, for the honour of the regiment and the flag that were all the patriotism they knew.

SWEDEN
1. Musketeer, 1709.—2. Sergeant of the Östgöta Regiment, 1757.—3. Drummer of the Södermanland Regiment, 1757. —4. Grenadier of the Södermanland Regiment, 1757. On active service the Prussian-style mitre was covered, to avoid any possible confusion with the enemy of the day.—6. Officer of the Jönköping Regiment, 1757.—6. Artilleryman, 1757.— 7. Yellow hussar.—8. Dragoon, 1757.—9. Blue Hussar. The sabretache had at its centre the monogram A.F. (Augustus Frederick), with crown. Note the incongruous effect of these uniforms in the midst of the extremely soberly dressed Swedish army.—10. Musketeer of the Kronobergs Regiment, 1788.—11. Trooper of the Östgöta Regiment, 1788. It is of interest to note the failure of all attempts to make concessions to contemporary fashion, which always resulted in hybrid, more or less comic-opera uniforms. These Swedish troopers can be classed with the Russian Guards, the top-hatted British soldiers, and some units of the Belgian patriot army.

Index

The page numbers in italic refer to the illustrations

Albermarle, Duke of *see* Monck
Album de la Caballeria española 126
Almansa, battle of 124
Alton, Comte d' 116, 118
Anjou, Philip Duke of 124
Arberg, Count Nicolas of 108
Ariès, Christian 28
Artillery, Austrian 114, 116, *117*
Artillery, British 43, 61–2, *63*
Artillery, French 10, 40, *41*, 42
Artillery, Prussian 80, *81*
Artillery, Russian 146, 152, *153*
Artois, Comte d' 12, 14
Augustus II of Poland 82
Augustus II of Saxony 42
Augustus III of Poland 82
Austerlitz, battle of 88
Austrian Succession, War of 26, 40, 82, 86, 102, 122, 126

Baillet-Latour, Count of 108
Bavaria, Elector of 32
Bavarian Succession, War of 66, 84, 86
Belfort, Comte Drummont de 17
Bercheny, Ignace-Stanislas de 36
Berdych 130
Binning, Thomas 62
Bland, Colonel 50
Blenheim, battle of 48, 86
Blücher 68
Boudriot, Jean 28
Bourbon, Philip V of 98
Breslau, battle of 102
Breslau, Treaty of 84
Brissac, Duc de 22
British Army of 1914, The 44
Bronikowski 64
Buturlin, Prince 132

Carabineers, French 17, 22

Carabineers, Russian 142
Catherine I of Russia 132, 134
Catherine II of Russia 128, 138, 144, 146, 148
Cavalerie française, La 24
Cavalry, Austrian 106
Cavalry, British 43
Cavalry, Freikorps 78, *79*
Cavalry, French 10, *11*, 12, *13*, 14, *15*, *17*, 18, *19*, 20, *21*
Cavalry, Household 44
Cavalry, Light 52
Cavalry, Line 48, *49*, 50, *51*
Cavalry, Russian 134, 140, *141*, 142, *143*
Cavalry, Spanish 126
Charles II of England 43, 46
Charles II of Spain 124
Charles III of Spain 124
Charles VI of Austria 86
Charles VI of Spain 126
Charles VII of Austria 86
Charles IX of France 10
Charles XII of Sweden 128, 150, 152
Charles Albert of Bavaria 86
Charles Emmanuel III of Savoy 122
Charles Theodor of Bavaria 86
Charrie, P. 26
Chartres, Duc de *see* Orléans, Philip of
Chasseurs, Austrian 94
Chasseurs, Mounted 26, 30, *31*
Chenier, Andre 76
Choiseul, Duc de 12, 17, 24, 42
Choppin, Henri 24
Clothing Book of 1742 44
Clouards, Condé de 126
Condé, Prince de 10, 18
Corneberg *see* Kronenberg
Cossacks 138, 144, 146, 150, *151*
Crook, Unton 43
Cuirassiers, Austrian 108, *109*, 110, *111*
Cuirassiers, French 20
Cuirassiers, Prussian 74, *75*, 76, *77*
Cumberland, Duke of 52

Danube campaign 48

Daun, Count 108
David, M. 36
Déak, Paul 32
Dessins des uniformes des troupes II. et RR. de l'annee 1762 102
Dettingen, battle of 50
Die osterreichische Armee von 1700 bis 1867 102
Discipline of the Light Horse 52
Dragoon Grenadiers, Russian 140
Dragoon Guards, 44, 46, 48, *49*
Dragoons, Austrian 104, *105*, 106, *107*
Dragoons, Belgian 108
Dragoons, British 48–50, 54, *55*, 56, *57*
Dragoons, French 22, *23*, 24, *25*, 26, *27*, 28, *29*
Dragoons, Light 54, 56, *58–9*, 60, *61*
Dragoons, Prussian 70–74
Dragoons, Russian 140

École de cavalerie, L' 17
Éléments de cavaleria 17
Elisabeth Petrovna of Russia 134
Engineers, Austrian 116
Engineers, British 62, *63*
Engineers, Prussian 80
Engineers, Russian 152, *153*
Épernon, Duc d' 10
Essai sur la cavalerie légère 17
Esterhazy, Count Ladislas-Ignace 36
Esterhazy, Georges 30
Esterhazy, Valentin-Ladislas d' 36
Eszlary, Charles d' 30

Feltz, de 32
Filtz, de *see* Feltz, de
Fischer, Jean-Chretien 26
Fontenoy 40, 48
Francis I of Austria 124
Frederick II of Prussia 34, 40, 64, 68, 74, 80, 82, 84, 86, 138
Frederick Augustus I of Saxony 82
Frederick Augustus II of Saxony 82
Frederick Augustus III of Saxony 84
Frederick the Great of Prussia 68, 80, 108

Frederick William I of Prussia 64, 65, 68, 70, 74, 80
Frederick William II of Prussia 68, 72, 76, 80
Frederick William III of Prussia 80, 82
Freiberg, battle of 76
Friedlingen, battle of 124
Frontier Guards 98, *99*
Fusiliers, Austrian 92
Fusiliers, Royal 61

Gassion 36
Golovtsino, battle of 150
Gomiecourt, Guillaume de 24
Gomiecourt, Raoul de 24
Gordon, Patrick 132
Görlitz, battle of 102
Goya 126
Granby, Marquis of 44
Grenadiers, Austrian 92
Grenadiers, Horse 43, 44, *45*, 46, *47*
Gribeauval, Jean-Baptiste Vaquette de 42
Guérinière, François Robichon de la 17
Gustavus IV of Sweden 152

Habsburg, Charles of 98
Handbuch der Uniformkunde 104
Hale 60
Hawley, General 52
Hinde, Captain R. 52
Höchstadt, battle of 124
Hohenfriedeberg, battle of 66, 76
Hohenkirch, battle of 102
Horse Artillery 42
Horse Guards, Dutch 44
Horse Guards, Royal 43
Horse Guards, Russian 134
Humbert 34
Hussars, Austrian 112, *113*
Hussars, 'Death's Head' 70, 126
Hussars, French 28, 32–37
Hussars, Prussian 64–7
Hussars, Russian 142, 146–9

Independence, American War of 14, 126

Infantry, Austrian 92–9, 100–3
Infantry, Hungarian 14, 96, *97*
Infantry, Russian 132–9
Infantry, Spanish 124–5
Ivan the Terrible 128

James II of England 43, 46, 48, 62
Jemappes, battle of 42
Joseph I of Spain 126
Joseph II of Austria 116, 120

Katholisch-Hennersdorf, battle of 66
Kellermann 38
Khovanski 130
Kingston, Duke of 52
Kleist, Colonel Friedrich Wilhelm von 78
Kloppelkrieg 120
Knötel, H. 104
Koehler 118, 120
Kolin, battle of 76, 102, 108
Kronenberg 32

Lacolonie, Marshal de 36
La Fayette 42
Lancers, Russian 142
Landshut, battle of 102, 108
Lawfeld, battle of 102
Leopold II of Austria 120
Lesczynski, Stanislas 82
Leuthen, battle of 66, 102
Lewenhaupt, battle of 150, 152
Liberation, War of 82
Liegnitz, battle of 102
Lienhart 34
Liessnaya, battle of 150
Life Guards 43–7
Light Cavalry, Russian 144
Light Horse, French 26, 30, *31*
Light of the Art of Gunnery, A 62
Light Troops, French 36, *37*, 40
Ligne, Field Marshal Prince Ferdinand de 108
Line Cavalry, Russian 140
Line Infantry, Russian 136
Lorraine, Francis of 94
Louis XII 10
Louis XIII 30

Louis XIV 10, 17, 20, 30, 32, 98
Louis XV 10, 17, 18, 20, 22, 24, 36, 126
Louis XVI 12, 14, 22, 36
Louis XVIII 22
Louvois 30
Löwenwald, von 132
Lunéville, Treaty of 102
Lützen, battle of 82
Luxembourg, Marshal François-Henri de 30
Luzzara, battle of 124

Malplaquet 48, 124
Manstein, Colonel Christoph Hermann von 132
Maria Theresa of Austria 86, 98, 114, 116, 122
Marlborough, John Churchill, first Duke of 61, 126
Massow, Colonel von 68
Maxen, battle of 102
Maximilian II Emmanuel of Bavaria 86
Maximilian Joseph III of Bavaria 86
Maximilian Joseph IV of Bavaria 88
Mazeppa 150
Medici, Marie de 22
Meersch, Colonel Jean-André van der 118
Menzel 68
Monck, George 43
Money Barnes, Major R. 44
Monmouth, Duke of 46
Monteils, Comte de 32
Morier, David 60
Mortany, Jacques-André de 32
Murray, Count 116

Namur, Congress of 118
Napoleon 38, 76, 84, 102, 120
Ney 38
Nicholas I of Russia 144

Oprichnina 128
Orléans, Philip of 36
Ottenfeld 102
Oudenaarde, battle of 48, 124
Oxford, Earl of 43

157

Parrocel 17
Patriots, Belgian 116–21
Paul I of Russia 134, 136, 138, 148
'Peasants' War' 120
Peter the Great of Russia 128, 130, 132, 144, 146, 150
Peter III of Russia 134, 138, 146
Peterwardein, battle of 72
Philip V of Spain 32, 122, 126
Philippe-Égalité *see* Orléans, Philip of
Plumet, le 26
Pokalem 24, *25*, 136
Polish Succession, War of 40, 102
Polleresky 36
Potemkin, Prince 138
Pragmatic Sanction 86
Prague, battle of 66, 102
Principles of Gunnery 62
Provence, Comte de 22

Rakoczy, Prince Francis II 68, 114
Ramillies, battle of 116, 124
Rastatt, Treaty of 122
Rattzki, 'Baron' de 32
Reichenberg, battle of 66
Reichstadt, Duke of 102
Reynolds 38
Rigondaud, Albert 26
Robin, Benjamin 62
Rocoux, battle of 102
Romonadovski, Prince 132
Rossbach, battle of 34, 40, 72, 76, 84
Rousillon 32

Ryswick, Treaty of 32

St-Cloud, M. Fouré de 26
Saint-Geniès, Marquis de 32
Saint-Germain, Comte de 12, 14, 17, 20, 40
Saint-Ignon, Count of 108
Savoy, Duke of 122
Saxe, Marshal de 42
Saxony, Prince Elector of 74
Scharawades 70
Scharnhorst, Gerhard David 82
'Schomberg' helmet 24
Schönfeld, Nicolas-Henri de 118, 120
Schony, von 78
Seven Years' War 34, 40, 42, 44, 48, 72, 78, 80, 82, 86, 94, 102, 106, 114, 122, 126, 152
Seydlitz-Kurbach, Friederich-Wilhelm von 17, 76
Shrapnel, Henry 62
Silesian War, First 66
Silesian War, Second 66
Silesian War, Third 66 (*see also* Seven Years' War)
Sofia of Russia 128, 130
Spanish Succession, War of 86, 124
Standards, British 52, *53*
Standards, French 18
Streltsy 128, *129*

Teuber 102
Thiennes, François Florent Count of 108

Torgau, battle of 66, 108
Trauttmannsdorf, Comte de 116
Trenck, Baron von 76
Turenne, Vicomte de 10, 17, 18
Turkheim, battle of 102

Uhlans, Austrian 114, *115*
Uhlans, Prussian 80
Ulm, battle of 88
Union, Act of 43
Utrecht, Treaty of 122

Vallière, Jean-Florent de 42
Verseilles, Marquis de 32
Vigo, battle of 124
Villaviciosa, battle of 126
Villers-Masbourg, de 118
Vincent, Baron 110
Voltaire 150

Wallace Collection 38
Walloon Guards 126
Walloon Regiments 98
Warburg, battle of 44
Waterloo 43
Westerloo, Marquis of 108
William III of England 44
Wolfe, General 60

York, Duke of *see* James II of England

Zieten, Hans Joachim von 17, 64, 66, 72, 80, 108
Zweibrucken, Charles Duke of 86